Sketches in Quantitative Finance

A Translation of Bachelier's
Le Jeu, la Chance et le Hasard

By Edward Harding

Kismet Press | MMXVII

Louis Jean-Baptiste Alphonse Bachelier (1870–1946)

Sketches in Quantitative Finance

A Translation of Bachelier's Le Jeu, la Chance et le Hasard

Edward Harding

kısmet•press

Libera Scientia | Free Knowledge

Sketches in Quantitative Finance
A Translation of Bachelier's Le Jeu, la Chance et le Hasard
by Edward Harding

Published in 2017
by Kismet Press LLP
15 Queen Square, Leeds, LS2 8AJ, UK
kismet.press
kismet@kismet.press

Translations, 1

Downloadable .epub edition also available

Printed and bound by IngramSpark with acid-free paper, using a print-on-demand model with printers in the US, EU, and Australia

A catalogue record for this book is available from the British Library

ISBN 978-0-9956717-8-2 (hbk)
ISBN 978-0-9956717-9-9 (ebk)

First edition printed in 2016
by Edward Harding & Venture Print Unlimited, Inc.
Plymouth, New Hampshire USA 03264
ISBN 978-1-944781-75-0

—§—

Contents

—§—

PREFACE

The Genesis: Students of finance (not the younger sleepy credit-baggers, but the older weathered ones who strive to understand how financial markets really work) tend to scour the literature for the answers, for the universal truths. These latter students may be motivated by the desire for wealth, fame, tenure, love or some combination of those human forces that drive us to do whatever it is that we do; those were the forces that led this author to seek and to discover Bachelier's work and to translate his *Le Jeu, La Chance et le Hasard.*

The First Challenge: My command of the French language is tenuous at best. But Bachelier's style of writing is straightforward, concise, consistent, repetitive, and easy to read, and thus it became a relatively uncomplicated work to translate.

The Second Challenge: The archaic financial instruments that Bachelier analyzes are structurally and contractually different than what investors encounter in financial markets today. Nonetheless, the theories and principles that Bachelier attributes to his financial instruments are readily transferable to contemporary world markets and contracts. In the end, his perspectives on the impact of randomness have all the characteristics of universal truths.

Disclosure: As Bachelier digs into patterns of valuation and randomness, he frequently discusses the difference between two values, for example the difference between an expected value, A, and an actual observed value, B. In his original text, the French word for this difference is *écart,* and in the first chapters of my translation, the English word *difference* seemed perfectly appropriate. As the text continues, these *differences* bear Bachelier's pedantic scrutiny, even more than the values A and B, but the significance of these differences seemed to become diluted when translated into English.

My slow realization of this dilution of significance led me to switch horses mid-text, and eventually to leave *écart* in its original French. It's worth noting that Bachelier, too, switched horses with his *écarts*: He expounds upon different *écarts* that have an equal probability of happening, logically named *isoprobable écarts.* But the cumbersomeness of, and frequent reference to, *isoprobable écarts* led Bachelier to add a note dropping the adjective "*isoprobable*" on the grounds that it should hereinafter be assumed.

Disclaimer: Bachelier refers to the work of a Dr. Gustave Le Bon as being particularly effective in the visual display of quantitative information. Unfortunately, there exists an unspoken moral ugliness in the nature of the displayed data (gender, race, brain size, intelligence) and in the provocative and offensive implications of Le Bon's work. But Bachelier is clearly interested only in the effectiveness of the displays of the statistics and shows no regard to Le Bon's underlying subject matter. My sincerest hope is that the reader of this translation will find Bachelier (and this translator) blameless in this matter.

Acknowledgments: This translation would not have been possible without the inspiration sparked by Louis Bachelier's original work. He deserves my greatest appreciation - and my apologies for mangling any of his thoughts. Nor would I have been able to complete this sometimes seemingly quixotic project without the life-support provided by my wife, Laurie; she deserves many times more recognition than she ever gets. My thanks also go to Christin Wixson for pointing me to Kismet Press in the first place and to Patty White for help in untangling some bewildering French. A special thanks to Tim Barnwell who took a personal risk to support this translation and who spent hours of editing to bring it to its final form. Finally, I have only myself to thank for all of the errors in translation.

EH 8 September 2017

—§—

Introduction

Louis Bachelier is hardly a well-known name in contemporary popular culture. Nor is "quantitative finance" a mainstream area of interest — and probably never will be. Nonetheless, there are currently some burning hot-spots in both academia and in the professional world of finance with people who are fascinated by the models that Bachelier proposed in the early twentieth century. Bachelier's work appears to have been the first recorded treatise applying probability theory to finance, although his work was only marginally recognized when he introduced it. Now, however, over a hundred years after publication, his work is celebrated as the product of genius, recognized for its originality, its breath, its technical sophistication and its relevance to contemporary quantitative finance.

In 1900, Bachelier published a text, *Théorie de la Spéculation*, based on his doctoral dissertation of the same name. In these works he proposes an original mathematical equation, a quantitative model, which describes the patterns traced by movements of a body in random motion. He proceeds to relate these random movements and some classical probability theory to yield a monetary value of municipal bond options.

The significance of Bachelier's model is well documented. Several Nobel Prizes were subsequently awarded to economists who borrowed, rejuvenated, expanded upon, or otherwise recognize Bachelier's original work as being the first to mathematically define many of the inherent concepts reflected in their much later models. Included among these later models are Fischer Black, Myron Scholes and Robert Merton's *Black Scholes Option Pricing Model*, Eugene Fama's *Efficient Market Hypothesis*, and Harry Markowitz's *Modern Portfolio Theory*. Mathematicians Norbert Wiener and Albert Einstein independently quantified *Brownian Motion* years after

Bachelier had already quietly done the same. Recently, contemporary authors in the field of finance give well-deserved recognition to Bachelier and to the significance of his contribution; a few examples include Benoit Mandelbrot and Richard Hudson's *The (Mis)Behavior of Markets* (Basic Books, 2004); Mark Davis and Alison Etheridge's *Louis Bachelier's Theory of Speculation* (Princeton University Press, 2006); and Justin Fox's *The Myth of the Rational Market* (Harper Collins, 2009).

The original *Théorie de la Spéculation* remains the flagship of Bachelier's work. But he also published several other books on the related topics of probability and speculation. One of these, *Le Jeu, la Chance et le Hasard* (Ernest Flammarion, 1914), [On Games, Luck (or chance) and Randomness] sold over 6,000 copies in its day. It is more approachable than *Théorie de la Spéculation* as it relates to popular games of chance (e.g. Poker, dice, Roulette), and discusses the qualitative nature of randomness in non-technical terms. But *Le Jeu…* also borrows heavily from Bachelier's original work and delves into financial speculation, especially into the valuation of the popular French municipal bond options. This book might be described as a historically significant introduction to probability theory's relevance in life's many and diverse everyday games.

Given the previous popular appeal of *Le Jeu…* and the renewed regard for Bachelier's lasting contributions, it came as a surprise to this author that an English translation of *Le Jeu, la Chance et le Hasard* appeared to be non-existent, or at least effectively unavailable. The text in hand seeks to remedy that vacancy on the English-language bookshelf.

— Chapter I —

Randomness

The idea of probability is inherent to consciousness and appears inseparable from it. As a result all existence is organized, in its fashion, from the calculation of probabilities.

Our actions are constantly guided by the search for maximum pleasure and minimum suffering, we always endeavor to do that which seems to us to be *probably* the most advantageous, it is thus always the sense of probability that directs us.

Reasonable acts are guided by the sense of probability, impulsive acts are the same, and impulsions are only the result of the previous accumulation of probabilities.

The idea of probability necessarily exists with consciousness, and forms, as it were, a body with her, cannot be the monopoly of the human species and as one cannot deny to the plants even a kind of consciousness, one must admit that they ultimately act from the sense of probability.

When, in a high forest, a shrub tilts to enjoy the single ray of sunlight that can filter through the leaves, it has its way of calculating probabilities, it is also obeying the principle of maximum pleasure, according to the law of large numbers and averages; it is doing that which is probably for it the most advantageous.

It is not certain that, from another perspective, in a very different direction, a sunny opening does not occur in the shade; but this is a very unlikely hypothesis, the shrub seems to know, it acts as if it knew.

By heading towards the clearing its tiny understanding chose the most probable hypothesis, it obeyed the principle of maximum expectancy.

While saying that the idea of probability is inherent to consciousness, that it is inseparable and that consciousness was born with it, we do not, fortunately, have to return to the sources of consciousness and to the causes from which it is derived, we do not have to solve the problem of birth and the evolution of instinct, or intelligence, or feelings of passions, we do not have to undertake the studies that go beyond our area of interest and were brilliantly exposed in other volumes of this series, we have only to find a result: the idea of probability always exists with the consciousness.

The ultimate degree of consciousness would be certainty, it is a limit to which one can approach without ever attaining, and mathematics itself is not absolutely certain, and strictly speaking, certainty does not exist.

If the sense of probability dominates all our actions, why has the calculation of probabilities come so late in science? For two reasons, that one can easily conceive.

The concepts of space and motion seem to us imposed by our senses, the notion of probability does not correspond to anything that is visible or tangible; I am perhaps the only one who has tried, not precisely to *visualize* probability, but at least to assimilate in certain cases these transformations to some physical phenomena; yet these phenomena don't present anything actually visible, one can only conceive them.

Another reason for the absence of any external manifestation or even any clear picture has delayed the birth of the calculation of probabilities. Another reason, besides having a bit of the same origin, has equally contributed to this result: The idea of probability seems in complete contradiction with the essential idea of exact science.

Exact science searches for the absolute, its principle appears to be the antithesis of the idea of probability. Likewise, the expression of calculating probabilities seems to imply a contradiction between those terms, as does also the analogous expression: laws of randomness.

It has required a great effort to dare to apply the mathematics which has the same precision to a question which, by its nature seems quite rebellious.

Huygens, the first to write a treatise on the subject, did not hide his enthusiasm. He said: "Nothing is more glorious than to be able to give rules to some things that, being dependent on chance, seem not to recognize any [rules] and thus to evade human reasoning." This phase has been repeated often and gives an idea of the interest which is attached to research the laws of randomness.

The factors that have delayed the emergence of calculating probabilities have had other consequences: some minds, even very fine, remain rebellious to this calculation, others will only painfully assimilate it, its acceptance will always be very difficult and its principles will perhaps always be misunderstood by the general public. On the other hand, for other minds, for those who have a tendency to meditation, for those who appreciate the philosophical studies as rational studies, for those who can finally understand, next to the beauty of a general law, the finesses of a subtle and delicate analysis sometimes brushing against the paradox, for those the theory of randomness will present an attraction and a special charm.

The calculation of probabilities is not, strictly speaking, an application of pure mathematics, it is an autonomous science having a special kind of aesthetic. To understand the beauty it is necessary to suppress the popular images that the consideration of a game engenders. Every manifestation of randomness may be considered as the result of a game and yet the dignity of the subject should not be diminished.

I do not have to describe the appearance of players around the roulette wheel, cartoonists have often addressed this easy subject. This is not exactly the same image that the same game evokes in the mathematician — he tries to imagine a sort of ideal fluid moving in an n-dimensional space while following the laws vaguely analogous to those of the movement of heat. This image is notably different, as we see, from the common concept of the game to its transcendent conception.

This is nearly always the game that can form an idea about a clear manifestation of chance, it is the game which gives birth to the theory of probability, it is to the game that this calculation owes its first stuttering as well as its latest developments, this is the game that allows the conception of this calculation in the most general way, it is thus the game that it is necessary to endeavor to understand, but it must be understood in a philosophical sense, regardless of any popular idea.

Can we define randomness?

"Strictly speaking, nothing depends on randomness; when one studies nature, one is soon convinced that its creator acted in a general and uniform manner, which bears the character of infinite wisdom and prescience. Thus to stick to this word 'randomness' an idea which must conform to the true philosophy, one must think that all things are regulated according to certain laws, of which most often, the order is not known by us, they depend on the randomness whose natural cause is hidden from us. By this definition one could say that the life of man is a game where randomness reigns." (De Montmort: *Essay d'analyse sur les jeux de hasard*, 1708) [*Essay on Analysis of Games of Chance*]

"All events, even those who by their pettiness appear not to adhere to the great laws of nature, are as a result as necessary as the revolutions of the sun. In the ignorance of the ties that bind the entire system of the universe, they were made to depend on final causes or chance, depending on whether they were happening or retreating with no apparent order or regularity; but these imaginary causes have been successively remote from the extent of our knowledge, and disappear entirely before sound philosophy, which only sees them as an expression of our ignorance of the true causes.

"Actual events have a link with previous events based on the obvious principle that one thing cannot begin to exist without a cause that produces it. This axiom, known as the 'principle of sufficient reason,' extends even to actions one judges as inconsequential. Even the most free will cannot, without a motive, give birth to these events; because if, all circumstances of two scenarios being exactly the same, she acted in one and refrained from acting in the other, her choice would be an effect without a cause; it would thus be, says Leibniz, blind chance of taste. The contrary view is an illusion of the mind that, losing sight of the fleeting reasons of the choice of the will in the inconsequential things, convinces itself that the event is determined by itself and without motive.

"We must therefore consider the present state of the universe as the effect of its previous state and as the cause of that which will follow. An intelligence that, for a given instant, knows all the forces by which nature is animated and the respective situation of the beings who compose it, if moreover it were vast enough to submit these data to analysis, would embrace in the same formula the movements of the greatest bodies of the universe and those of the lightest atom:

nothing would be uncertain for it, and the future, like the past, would be apparent to its eyes." (Laplace: *Essai philosophique sur les probabilités*, 1814) [*Philosophical Essay on Probabilities*]

"How dare mention the laws of randomness? Isn't randomness the antithesis of all law? In offering this definition, I will not propose any other. On a subject loosely defined, one could reason without equivocation. Is it necessary to distract the chemist from his burners to press him on the essence of matter? Does one commence the study of the transmission of power by defining electricity?

"The word 'randomness,' intelligible in itself, awakens in the mind a very clear idea." (J. Bertrand: *Calcul des probabilités*, 1889) [*Calculation of Probabilities*]

"First of all, what is randomness? The Ancients distinguished the phenomena which seemed to obey harmonious laws, established once and for all, and those that they attributed to chance; it was those phenomena that they could not predict because they were rebellious to any law. In each area, specific laws do not decide everything, they just traced the limits within which they were allowed to move randomly. In this design, the word randomness had a precise objective meaning: that which was random for one, was also random for the other and even for the gods.

"But that concept is no longer ours; we have become absolutely deterministic. And even those who would set aside the laws of human free will leave at least determinism to prevail, without sharing, in the inorganic world. Every phenomenon, however small it may be, has a cause, and a mind infinitely powerful, infinitely knowledgeable of the laws of nature, had the power to predict since the beginning of time. If such a mind existed, we could not play at any game of chance with him; we would always lose.

"To Him, indeed, the word random would not have any meaning, or rather, there is no such thing as 'randomness.' It is because of our weakness and our ignorance that there would be a word for us." (Henri Poincaré: *Revue du mois*, 1907 — *Science et Méthode*, 1908) [*Review of the Month*, *Science and Method*]

Thus one cannot properly define randomness for this reason; that in reality, randomness does not exist.

We say that a phenomenon is due to random chance when its causes are unknown to us and appear to be beyond analysis.

The last part of this kind of definition is necessary; we would not consider that an event was due to randomness when the causes that produce it, by being unknown, appear to us to be simple.

We are ignorant of the causes of most of the facts, we only attribute one part to randomness.

One easily sees that the definition applies to the facts that one usually considers to be coincidental. The coin thrown into the air swirling, falls on the side of heads or the side of tails, according to randomness; it is randomness which turns over a selected card in a game, or which produces an unexpected encounter with a person who is rarely seen, etc.

If one knows the order in which the cards from a deck are arranged, the turn of a certain card does not depend on chance. If one shuffles the cards for a long time, the causes that can produce turning of a designated card become unanalyzable, the feat is dependent on chance.

The forces that one gives to the coin by tossing it in the air depend on many small causes that seem beyond analysis. If one assumes the habit of always tossing the coin in the same way, the landing of tails does not depend on chance, the causes of the forces would seem to us analyzable.

The chance meeting of someone who is rarely seen is attributed to randomness because the causes of the meeting appear unanalyzable. If the person passes every day where the encounter occurred, we do not attribute it to chance because the causes are simple.

The number of facts that we attribute to randomness, fictitious causes created by our ignorance, must vary like that ignorance, depending on the times and on the individuals. That which is random for the ignorant is not necessarily random for the scientist. That which is random today may no longer be tomorrow.

Scientific discoveries can restrict the scope of randomness, since they narrow our ignorance. There are, however, phenomena that we would consider as before, that always depend on randomness because we cannot perceive the possibility of some simple cause that can replace it. The picking of a certain card from a well-shuffled deck depends on randomness and we seem to always count on it.

On the other hand, there are phenomena that we attribute only provisionally as random:

For a hundred years we thought that before long meteorology would become a real science permitting us to predict the exact weather into a certain future; it was thought that weather prediction would soon pass from the domain of randomness to that of certain knowledge. We still believe that today, but perhaps with less faith. The causes that produce the atmospheric changes are minimal and many, they don't necessarily have uniform results from which would arise great laws, so it is quite possible that randomness, for a long time, will seem to govern the effects.

We cannot, in this little book, study all the manifestations of randomness that are possible to submit to calculation and to process in a scientific way; thus we will study only very superficially the applications of the calculation of probability theory to biometric statistics. This question presents a very great interest, but it alone would require a new book. By confining ourselves to generalities, analyzing games, speculation, some observational errors, we would already have to cover quite a large area of understanding the fundamentals of the calculation of probability.

That word calculation must not alarm, we will not do any calculations here and our study will be only descriptive.

— Chapter II —

Probability

One defines the probability of an event as the ratio of the number of favorable cases of the happening of this event to the total number of possible cases.

The probability of throwing four pips, for example, with one die is 1/6 because six cases may present themselves when the die is rolled on the carpet and only one is favorable to the showing of four pips.

The probability of turning a king on a deck of 32 cards is 1/8: there are in fact 32 possible cases and four favorable cases; the probability is 4/32 = 1/8.

If an urn contains a white ball and two black balls, the probability of drawing a white ball is 1/3, the probability of drawing a black ball is 2/3.

If one tosses a coin twice, the probability of once getting heads and once getting tails is 1/2. In fact, four cases are possible: one could get heads twice or tails twice, or heads the first toss and tails the second or tails the first toss and heads the second. These last two cases are likely, the probability is thus 2/4 or 1/2.

The definition of the probability always assumes that the cases are equally likely.

In the first example given above, one should be careful saying: "The die can show the side with four pips or it can show another side; thus there are two possible cases with one favorable: the probability is 1/2." The two possible cases are not equally likely.

In the fourth example, one should not say either: "one could toss heads two times, or tails two times or heads one time and tails one time, there are three possible cases and one likely, the probability is 1/3." The three possible cases are not equally likely.

From the point of view of mathematics: the division of cases of equal probability is a result of the studied problem.

The consequences that the calculation of probabilities deduced from these data are mathematically exact.

One tosses a coin at random, we admit that it is as likely to land on tails as heads, that is a given of the problem.

Once this given is accepted, the probability of obtaining tails twice in a row is 1/4; that result is mathematically correct for the same reason that two and two makes four.

The probability of obtaining in three tosses twice heads and once tails is 3/8, this is an absolutely necessary consequence of the givens.

If one wants to experiment and if the outcomes are not exact, it would not disprove the calculation of the probabilities such that the results would appear erroneous; the calculations only faithfully translate the hypothesis. If the coin is unsymmetrical or if those who toss it knows how to favor the appearance of the side that he desires, it is evident that the calculations could not predict the outcome.

That remark would appear naive if it were any other math question, one knows very well that there is in a calculation only that which one puts there. When it comes to probabilities, the imprecision of the subject can make us lose sight of this basic truth.

One immediate consequence of the definition of probability is that the value is always between zero and one, the latter value corresponding to certitude, the former to impossibility.

Other immediate consequences: the sum of the probabilities of every possible case is equal to one.

If one designates by the letter p the probability of an event, the probability of it not happening (or, as one says, of the contrary event), is 1-p.

In ordinary language, one often uses the word *chance* in place of the word *probability*, thus one says of an event that it has nine chances in 10 of happening by expressing that the probability is .90 or 90%.

At the beginning of the calculation of probabilities, one uses the word chances and also the word randomness. One designates the probability, likewise in the algebraic formulas, by a fraction in which the numerator represents the number of favorable chances and the denominator the total number of chances. Today, in most formulas, one designates the probability by a single letter.

— Chapter III —

Appreciation of Probabilities

Let's imagine a mind that can exactly know probabilities.

This mind would not be comparable to that assumed by Montmort, Laplace and Poincaré, for whom there existed only certainty. Randomness would exist for the mind that we are considering, but he would make the best of it; he would be only a half god but he would be very superior to us. With him one could, without any thought, play at a game of chance, at dice for example, but he would be imprudent to play at cards. He wouldn't win consistently, he wouldn't be able to put in his cards the trumps that chance might have denied to him but, at a fair game, he would have a great superiority and would necessarily finish by winning.

In life it would be the same; fate could be unfavorable toward him, it would at least not be very advantageous on the whole.

Without possessing the absolute science, if we could fully appreciate probabilities while abandoning to chance for the most part that which our weakness must give up, we would certainly be much stronger.

To speak of the poor appreciation of probabilities, that is to put on trial the ignorance of humanity, the subject is infinitely vast and we must restrict ourselves to a general idea; the weakness of our intelligence has the effect of making us consider facts as random or nearly certain, or nearly impossible.

We have a natural tendency to embrace the probabilities by their extreme values of zero and one.

For the inferior minds, it seems that it would be difficult to consider the total range of probabilities, the believable and the completely unbelievable are nearly the only two alternatives between which a mediocre mind can hardly balance. Public opinion is generally formed from extreme ideas equally poorly based.

I find it interesting to reproduce verbatim that which was written on the subject by Montmort and Laplace, two masters on the issue.

"Everyone knows that due to the failure of the evidence we must look for some likelihood in order to find the truth; but we don't quite know whether there are some greater likelihoods and lesser ones, to infinity and for the mind to be a good judge, one must distinguish all the degrees" (de Montmort).

"The mind has these illusions like a point of view, and just as touching them corrects them, the reflection and the calculations correct the first.

"Our passions, our prejudice and the opinions dominate, and exaggerating the probabilities which are favorable to them and discounting the contrary probabilities are some abundant sources of dangerous illusions.

"The present evils and the cause which gives them birth, influences us much more than the memory of the ills produced by the contrary cause; they prevent us from appreciating with fairness the inconvenience of one from the others and the probability of the usual means save us from them. This is what leads the people alternately towards despotism and toward anarchy leaving the state of rest, in which they only ever return after long and cruel agitations.

"That vivid impression that we receive from the presence of the events, and which leave us to hardly mention the contrary events observed by others, is a principle cause of error, from which we cannot too much insure ourselves....

"One doesn't consider at all the great number of non-coincidences which don't make any impression or that one ignores. Meanwhile it's necessary to understand them in order to appreciate the probability of the causes to which one attributes these coincidences" (Laplace).

If you have an instinctive tendency to bring closer the probabilities from their extreme values, this is due to the emotions, to the beliefs, to the prejudice and also to a kind of habit which makes us assimilate in part the events which we would have to consider as accidental to those that we have ordinarily considered as certain or as impossible.

— Chapter IV —

Mathematical Expectation

Probability is not the only quantity that is interesting to study in the questions related to randomness; it is very different to have one chance in ten to win 100 francs or one chance in ten to win 1 million; it is that consideration that leads to the notion of mathematical expectation [expected value].

One calls mathematical expectation of an eventual benefit the product of that benefit multiplied by the probability of realizing it.

If you have one chance in two of winning 2 francs, the mathematical expectation is 1 franc.

In a lottery there is only one prize of 1 million francs and two million tickets, one possesses a ticket, whose value, that is to say the corresponding mathematical expectation is 1,000,000 multiplied by 1/2,000,000, or .50 francs.

One could say that the mathematical expectation is the value of a sum for which the gain is only possible.

As soon as it becomes a question of a loss instead of a gain, a loss could be considered as a negative gain, and the mathematical expectation is negative.

If one has one chance in 20 of losing 40 francs, the mathematical expectation is:

$$-40 \times 1/20, \text{ or } -2 \text{ francs.}$$

If there is a 1/100 probability to suffer a loss of 1,000 francs, the mathematical expectation is:

$$-1{,}000 \times 1/100, \text{ or } -10 \text{ francs.}$$

A negative mathematical expectation is the value of a loss for which the realization is only possible.

In the last example, one could again say that 10 francs is the premium that should be fairly paid in order to avoid the risk of losing 1,000 francs.

The mathematical expectation is thus the value of a possible gain (or of a possible loss).

The *total* mathematical expectation to a player is the sum of the products of the possible benefits multiplied by the corresponding probabilities.

If, for example, a player has one chance in three of winning 16 francs, two chances in three of losing 6 francs, and one chance in 12 of winning 20 francs, his mathematical expectation total is $16 \times 1/3 - 6 \times 2/3 + 20 \times 1/12$, or 3 francs. That sum of 3 francs is the value of his transaction, which contains three eventualities. It is against that sum of 3 francs that he could, without advantage nor prejudice, abandon that transaction.

In order to make the total mathematical expectation, it is simply a matter of adding the expectations which correspond to the gains, then subtracting those which correspond to the losses.

The mathematical expectations are the sum of fictitious amounts; it is necessary to understand by this that they don't ordinarily correspond to a possible value of gain or of loss.

If one has one chance in ten of winning 20 francs, the mathematical expectation is 2 francs; it isn't a possible value of gain, the possible values are 0 and 20 francs.

It is evident that the mathematical expectations can be added like the sums of ordinary amounts, the total mathematical expectation is the sum of the corresponding expectations of all the eventualities which can be realized. That property of addition often makes the calculation of mathematical expectations very easy; in most cases it is easier to calculate a total mathematical expectation than each of the terms of which it is composed.

It is evident that a game is advantageous when the total mathematical expectation is positive, that it is disadvantageous when the total expectation is negative, and that it is neither advantageous nor disadvantageous when the total expectation is null. One then says that the game is *equitable*.

If, for example, a player has one chance in three of winning 12 francs and two chances in three of losing 6 francs, his total mathematical expectation is null, the conditions of the game are neither advantageous or disadvantageous to him; the game is equitable.

If a game is composed of several hands, the total mathematical expectation is the sum of the respective expectations of the different hands that comprise the game (on the condition that all the hands will necessarily be played).

In particular, if the game is consistently identical to itself, the total expectation is proportional to the number of hands.

If a game is equitable at each hand, it is equitable in its entirety.

There doesn't exist any combination that can render a game advantageous or disadvantageous if the game is equitable at each hand.

In the simplest games, one first deposits a bid or stake and thereby acquires a positive mathematical expectation; the game is equitable when the stake is equal to this expectation.

This is a particular case of the general definition, the bid that one has given up is, in effect, a negative expectation equal to the bid.

The games played in the gambling houses and the casinos are not equitable; they are obviously disadvantaged for the players.

Even a very slight disadvantage will completely change the results of the game over time; we are going to have to pursue this question.

As soon as one says that a game is equitable, one must understand that the total mathematical expectation is zero, nothing more.

An equitable game can be unreasonable: two people each with exactly a million francs could play the game of heads or tails; the game is equitable, and yet such players would be considered fools.

On the other hand, one might risk a louis or two at the roulette wheel without deserving judicial counsel.

A game could be unfair from the mathematical point of view and be nonetheless very practical: one considers it very prudent to insure oneself, and meanwhile, from the mathematical point of view, insurance is unfair.

§ 1. — Game of Pass Ten

For three centuries, people have played the game of "pass ten." The rule is very simple; one rolls three dice at random; one of the players wins if he obtains a total of points greater than 10; he loses if the sum of points is less than or equal to 10.

One easily sees that the score of 11 has the same probability as the score 10, that the score 12 has the same probability as the score 9, etc. Thus the game is equitable.

A friend of Galilée is surprised to see the score of 9 happens less often than the score 10, but that each of these scores can be obtained from six different combinations.

Galilée makes the remark that the six combinations are not equivalent in the two cases. For example, the three dice give 3, 3 and 3, the score is equal to 9; if the dice give 3, 3, 4, the score is equal to 10; the two cases are not analogous: the combination 3, 3, 3 is unique, while there are three combinations of 3, 3, and 4; they are 3, 3, 4, 3, 4, 3 and 4, 3, 3.

Galilée shows that in rolling three dice he can produce $6 \times 6 \times 6 = 216$ different cases and that among those, 27 are favorable to the outcome of a score of 10, but that only 25 are favorable to the outcome of a score of 9.

Cardan and Galilée are considered as precursors of the calculation of probabilities.

§ 2. — Game of the Match

Another well-known game is the "game of the match." The banker, that is to say the one who holds the deck of cards, successively turns over 13 cards from the 52 cards of a complete deck while calling: ace, two, three, … jack, queen, king. There is a match when the number (or the face) called is the same as the number turned over. The banker gives a franc to each of the players and he receives from each of them as many francs as there are matches.

Two centuries ago, de Montmort, wanting to estimate the advantage to the banker, calculated the probability for which there would be only one match, for which there would be only two, for which there would be three, … for which there would be 13. He also saw that the banker was playing fairly, then he estimated the uselessness of those calculations (at least to resolve the proposed problem) and their result appeared to him as nearly evident.

Before beginning to play, the probability for which the banker would turn over a called card is 1/13. For example, the probability for which he turns a five while calling five is 1/13, because there are four fives in the 52 cards, and thus the probability is 4/52 or 1/13. Since there is one chance in 13 that the banker wins a franc while calling the five, his expectation for this round is 1/13. It is the same for each of the instances and as there are 13 of them, the total value is 13/13, that is to say — one.

The banker has thus paid 1 franc to each of the players and, as he acquires by the game an expectation of 1 franc for each of them, he plays fairly, his game is fair.

We have there a very curious example, becoming classic, of the case where the mathematical expectation is obtained very easily by a direct process, such that the probabilities related to the same problem are from a very laborious calculation.

The probability for which there would be, I suppose, exactly three matches, can be determined only by tight and delicate reasoning, however it differs with the number of decks of cards used; on the other hand, the mathematical expectation is obtained without any difficulty and is independent of the number of games, provided that one turns only 13 cards.

The game of horloge is analogous to the game of match, with the difference that the banker receives 1 franc from the players if there is one or more matches and that he pays them 1 franc if there isn't a single match.

The game is outrageously advantageous for the banker: the probability that he wins varies between 0.63 and 0.65, depending on the number of hands of cards.

§ 3. — Games of Cards, in General

One cannot conceive the idea of the prodigious number of different results that one can obtain with a game of cards, or even with a simple game of dominos.

The number of possible arrangements of 52 cards is expressed by an 8 followed by 67 other digits.

The number of possible arrangements of 32 cards is 263 million of billions of billions of billions.

Two players A and B each hold seven random cards from a deck of 52 cards; the number of possible hands is six million of billions.

Assuming that they play continually, one hand per minute and that their hands are different for each hand, they would only exhaust all the possible combinations at the end of one hundred million centuries.

One understands that it is generally impossible to study, in a complete fashion, a game of cards by the calculation of possibilities.

The interest in the game is not diminished by this. On the contrary, although the player cannot calculate the probabilities, he must make a decision based on his intuition and on his experience and to act as if he knew the value of it; the game no longer seems to him thus as uniquely dependent on randomness, his ability plays a certain role here.

By placing oneself in this point of view, one can divide the games into three categories: the games of pure luck (heads or tails, games of dice, some card games, roulette, Thirty and Forty, etc.), the games of luck and skill (most card games, games of dominos, etc.) and the games of pure skill (chess, checkers, etc.).

We will not study each game in particular, our goal is to make comprehensible the general theory which applies to any game, which can be can be applied to games forgotten today as well as those that could be invented tomorrow.

§ 4. — APPLICATION TO THE DETERMINATION OF PROBABILITIES

Like the application of the notion of mathematical expectation, we are going to solve a very simple problem and meanwhile out of interest and from considerable scope, we'll discuss the important question of the ruin of players.

Player A possesses *a* francs and player B possesses *b* francs, they play at heads or tails and, after each round, the loser pays 1 franc to the winner.

These two players play until one of them has lost the total of that which he possessed.

What is the probability that player A finally earns the b francs of player B?

In order to simplify the language, one calls the *fortune* of a player the total sum that he dedicates to the game; as soon as he has lost that sum, one says that he is ruined. Thus, the considered problem could be stated in the following fashion:

Player A has fortune a, and player B has fortune b, they play at heads or tails and, after each round, the loser pays 1 franc to the winner.

They play until the ruin of one of them. What is the probability that player A ruins his adversary?

To resolve the problem, simply use the notion of mathematical expectation: The game is fair, the mathematical expectation of player A is null.

No limit was assigned to the duration of the game, only two eventualities are possible: either player A finishes by winning the sum b possessed by his adversary or he himself loses the sum a that he possessed.

Let P be the probability of the first eventuality, the probability sought. Player A would have probability P to win the sum b, his expectation, for that eventuality, is Pb.

Player A also has probability (1 - P) of losing the sum a, his expectation for that eventuality is $-(1 - P)a$.

The total mathematical expectation is null, one has:

$P\,b - (1 - P)\,a = 0$, from which

$P = a / (a + b)$.

The probability that A wins is thus

$a / (a + b)$, the probability that B wins is, by the same reasoning,

$b / (a + b)$.

The probabilities of gain for the players are proportional to their fortunes.

The probability that A will be the final winner is the probability that B will be ruined, thus:

As long as two players play at an equitable game without establishing in advance a maximum number of hands, their probabilities of ruin are inversely proportional to their fortunes. It follows that the poorer will certainly be ruined.

While playing against a very rich adversary, the ruin is almost certain. Those who play against all adversaries who present themselves, find themselves in the same conditions as if he played against an adversary very rich: his ruin is almost certain.

The probability of his ruin approaches certainty while at the same time he believes his expected gain is indefinite.

This result is obvious anyway, the player wanting to win a big fortune with a little capital must, in all fairness, be almost certain of losing his little capital. If it were not thus, the player would clearly be favored.

Therefore, when a player plays fairly against all adversaries who present themselves, his ruin, in the long run is certain; he will be able for a while to win much, perhaps, but he will always finish by ruining himself.

The ruin is more certain when the game disfavors the player, but it is not the same when the conditions of the game favor him, however small the advantage is.

Then not only is the ruin no longer certain, but the player is nearly certain not to be ruined if, at each hand he risks only a small portion of his capital.

I will return later to this question of the ruin of players, one of the more interesting of our study.

— Chapter V —

Moral Expectation

Instead of considering an eventual benefit having an intrinsic value, Daniel Bernoulli had the idea of relating that benefit to the fortune of those who expected it.

A gain of a thousand francs matters little to a millionaire, it matters a lot to a pauper; the mathematical expectation, such as we have defined it, is no different between the two cases.

It is in order to distinguish them that Daniel Bernoulli proposed to add to the notion of mathematical expectation the notion of moral expectation.

He supposed that the eventual gain must be related to the fortune of those who expected it. A gain of a million to those who already possesses a million must, according to him, give the same "morale expectation" as a gain of a 100 francs to those who possess only 100 francs.

Based on that idea for which the exaggeration is evident, he demonstrated with figures to support [the notion] that the game considered to be fair is in reality disadvantageous, that insurance, by contrast, is advantageous, that one must not risk all one has in one aleatory enterprise; he demonstrated, in a word, a bunch of things that appear to be common sense but which his theory and especially his data only evoke weak arguments.

One result, meanwhile, is very interesting: Bernoulli noted that all these conclusions subsisted while supposing simply that the morale expectation of a benefit is especially less grand for those for whom the hope is greater riches.

In fairness to mathematical expectation, the moral expectation would not vary any more due to the reverse of fortune, it diminishes only with the fortune. It is evident that the function which expresses the influence of fortune was not specified, it could not be a question of obtaining data.

Reduced to that vague form, the idea of D. Bernoulli is very tenable and appears likewise very fair to the condition of the unique reporting of an individual considered in isolation.

It becomes untenable if one wants to report on many.

In the games of commerce or industry, 1,000 francs is worth 1,000 francs, whether the possessor is rich or poor. Each time that there is an exchange, a sum of money, that really exists or which is expected, assumes an intrinsic value. D. Bernoulli, anyway, agreed, he never tries to impose his hypothesis.

The theory of moral expectation has never been applied; it would condemn all the games and most of the aleatory transactions between persons of very different fortunes. One example is going to convince us of this.

Player A possesses a hundred pieces of gold and player B, a hundred pieces of silver. Player A, plays with equal chance of gaining or losing a piece of gold. One understands that the player A will not be so naïve as to play with B one piece of gold against one piece of silver for this simple reason: that gold is worth more.

In the hypothesis of D. Bernoulli, one distinguishes the *moral fortune* from the *physical fortune* or numerical fortune. The first which is related to the second by a logarithmic formula, is supposed to represent the pleasure which corresponds to the possession of the second.

D. Bernoulli was careful to note that strictly speaking the physical fortune is not equal to the numerical fortune, the latter can be null, the first is never; a man who doesn't possess a sou has meanwhile a capital represented by his physical, moral and intellectual qualities.

The theory of Daniel Bernoulli secured in his period, and even much later, a very great success to which Buffon, by his popularity, contributed for a large part. Nicolas Bernoulli, excellent mathematician, cousin of Daniel and nephew of Jacques, of whom we will discuss, did not welcome without reservation the theory of moral expectation and his enthusiasm was more tempered; he noted that if, in the conducting of one's life, one should follow from a distance the suggestions of this theory, it would be necessary to protect oneself from applying it to the determination of investments.

Daniel Bernoulli introduced the methods of infinitesimal calculus into the theory of probabilities, it is there a beautiful claim to fame that posterity will give him. In all his works, besides, he shows a profound and original spirit.

Cramer proposed another hypothesis. The moral fortune will be proportional to the square root of the numerical fortune.

A man four times as rich as another receives from his fortune only double the satisfaction.

That hypothesis is as simple and interesting as that of D. Bernoulli, but it is equally arbitrary and results in the same criticisms:

Or it is a matter of a game (or of an aleatory transaction) and the hypothesis is inadmissible since one must then uniquely consider the intrinsic value of a sum. Or it is a matter of the value that the possessor, that is to say an individual determined considered at an instant determined, attributes to this sum. The hypothesis is thus arbitrary and much too simple.

— Chapter VI —

The General Idea of Expectation

Let us suppose that it is no longer a question of a simple sum of money and that the object of desire is any honorific or sentimental material.

Let us say up front that this is the idea of expectation which is innate in us and not the idea of probability.

It is through an effort that, in an expectation, we are able to separate the two factors: the object of desire or fear on one hand and the probability of the realization on the other. In a natural state, the two factors are mixed together.

All civilized beings obey a principle of maximum expectation: they always act in a way which seems to them to be most advantageous. It is the expectation that guides them, it is not the probability which is only one factor.

Since the expectation directs all of our acts, it is not without interest to research until that point that expectation can be assimilated to a mathematical expectation, a simple product of two numbers.

In expectation, it seems at first that we can abstract the probability from the realization of the desire and to consider that expectation as composed of one part from the probability, the other part from something very poorly defined and very vague which will be the value of the pleasure before the result of the realization of the awaited fact, or rather the value that we actually attribute to it.

If we think about our hopes in a healthy way, we could perhaps, in certain cases, arrive at isolating the probabilities and at decomposing our expectations into two elements, of which one would be simple. In fact, there is none of that, in our imagination the probability of an expectation is related to the importance of the object of hope;

we exaggerate the probability of that which we desire ardently, we conclude, as the popular expression says, by assuming our desires to be realistic.

Without that exaggeration, without the illusions of which our imagination enjoys fondling, life would be perhaps untenable; it is most fortunate that our expectations are not too mathematical and that the probabilities that create our thinking are unanalyzable and hyper-connected.

When an expectation is negative, that is to say when it is a question of fear of a misfortune that one apprehends, of a danger that one dreads, it is rare that one exaggerates the importance of it. There is there again a reprieve, a benefit which aids living.

Among the different factors which influence our expectations, it is especially necessary to consider time. Time is on our minds in a curious fashion and a hope differs greatly from a mathematical expectation when time plays a grand role.

From the point of view of mathematics, the influence of time translates itself in a manner very simple.

One discounts with compound interest a mathematical expectation as a sum of money necessarily realizable; one obtains thus its actual value.

It is on that principle that the calculation of insurance is based. If one has one chance in 10 of winning 100,000 francs in five years, that expectation will be worth 10,000 in five years. In discounting that sum of 10,000 francs with compound interest, one obtains the actual value of the expectation.

That is so simple, but when it is a question, more generally, of a hope, our imagination greatly complicates the choices. It distains the geometric progressions and those discounting formulas are transcendental.

If the analytical functions that give birth to our thoughts are not expressible or algebraic or by any other manner, they nonetheless have all the same general allure held by those that are of human origin.

Just as we all have nearly the same skeleton and the same arterial system, likewise we analogously feel happy and painful impressions. The result is that the laws of relative discounting to our desires exhibit some general characteristics:

When the period of the possible realization of desire is too far away, we attribute only little value to hope. The brevity of life and the uncertainty of the future are maybe the causes of it.

The mathematical expectation of an eventual gain is constantly growing until that moment when fate decides if the gain is realized or not. At that instant, if fate is favorable, the mathematical expectation grows suddenly and acquires the value equal to the gain. If fate is unfavorable, it falls all of a sudden to zero.

That is not what usually occurs when it is a question of desire; very often, at the moment of the realization of desire, the dream fades and is replaced by disillusion. There is a fall of expectation even when all is favorable.

The period which ought to produce the happy event could be uncertain and that uncertainty can have a great influence on the expectation. In general, beyond a certain time, the desire loses its intensity; a pleasant event stretched out too long is often greeted with indifference.

In summary, one cannot with exactitude, equate a desire which is necessarily complex to a mathematical expectation which is infinitely simple.

It is not less true that mathematical expectation, therefore, precisely, from its simplicity, gives to the general idea of expectation a sort of very expressive and very clear analytical representation.

— Chapter VII —

Luck

What should we mean by "to have luck?" Could the calculation of the probabilities specify some vague little sense that we attach to these terms and is it possible to bring some clarity to the concept of an idea so poorly defined and so uncertain?

It is difficult to define this subject and it seems that mathematics cannot add great things to the inclinations of good sense; good sense banishes all mystical ideas or superstitions.

There are moments in life when the simple fact, or the tiniest event can have a considerable repercussion on the future wellbeing. At these instances, one has the very clear impression of being at the mercy of randomness, as if fate is to be decided by a throw of the dice. One has good luck when the randomness is favorable, as if it were a game.

At certain critical moments, the randomness appears to us in all its force, especially more frightening that it is elusive and that we are defenseless against it; but randomness always exists and certainly one can find a trace of it at each instant even in the smallest things.

An individual has luck when the random events which influence the course of his life are more favorable than they ordinarily are to him; he has luck in life as he would have in a game. The favorable random events will be thus equated to the gains, the unfavorable random events to the losses.

Just as in a game, one cannot win without interruption and in a continual fashion, by the same fortune, under these multiple manifestations, cannot constantly favor one same individual without some alternations and without some regression. The comparison of life as a game of chance seems for this reason, rather justified. But if the game gives a very simple image permitting the making of an idea of that which may be luck, it is not necessary, without some reserve, too push too far the assimilation: the luck of today contributes to the luck of tomorrow, the successive hands of the game are not independent and their dependence is too complex for us to be able to analyze. We must content ourselves with a whole idea: the actual luck seems to favor the future luck and the present adversity seems to foretell the misfortune of the future.

Thus a grand malady, weakening the organism for a long time, renders the future much more somber.

Let us take the individual at his birth: he is born with some physical and moral qualities of which the effect will endure all his life, this first round that he plays unconsciously has extreme importance, the fortune of all his existence depends on it.

This is the moment especially that he must have good luck.

At each instance, in the following events, he will play a new round, but his issues will be very unequal; sometimes he will make a huge gamble without hesitation, passing nearby a danger that he doesn't even suspect.

His ability, sometimes likewise his intelligence, will render the game more favorable, the luck of which would not lose its advantage.

That which dominates in the considered game, which is the general characteristic of it, is that the previous losses render more probable the actual losses; the differences have a tendency to become larger than they would be if luck alone was the cause of it.

In the general theory of probabilities, one limited oneself in the past to consider the games in which the successive rounds are independent, that is to say to study uniquely the effects of chance only. The case where it is a matter of chance alone is a limited case and the idea has come to me, several years ago, to study some of the cases where random events are concurrent with other causes, these causes can influence the effect of chance, as chance can have an influence on them.

I call these probabilities *related probabilities* which don't depend uniquely from chance and it has been possible for me to study several classes of these probabilities.

The domain of related probabilities is immense, it extends from pure chance to the absolute certainty.

The little that we can know about these probabilities permit meanwhile to specify, to a certain degree, the problem of luck:

In one of the chapters of my treatise on the *Calcul des probabilités* [*Calculation of Probabilities*], I study the case of a game where in each instance, the effect of chance depends on the results of the previous effects; a previous advantage renders more probable an advantage for the considered round, a previous loss renders more probable a loss.

This is the equivalent to what generally occurs in life: the luck today favors the luck tomorrow.

The theory would take into account how it could vary with time with that which remains of past effects; its formulas contain two arbitrary functions of time which give them a beautiful elegance and a grand generality, but that generality is only relative, it is entirely insufficient to represent the hazards of life.

In a following chapter of the same book, I study the case where the conditions of the game at each round depend on the maximum past loss. The equivalent occurs in life, it is rare that the moment when one has sustained the most grand troubles doesn't leave a lasting scar: a man disfigured in a train accident keeps an inferiority all his life.

The theory of connected probabilities does not evidently permit making a precise theory of luck, it permits meanwhile to ameliorate to a great measure that image, already classic, where life is compared to a game.

§ 1. — Happiness

The luck of an individual must be appreciated by others; as soon as it is appreciated by the person concerned himself or rather reported to him, the luck becomes happiness.

Happiness (unhappiness when it is negative) varies between more tightened limits than luck; the pauper has much less luck than the rich, he doesn't necessarily have much less happiness.

Happiness depends on luck, but from the laws which hold not only to the person considered, but also to an infinity of other causes.

If one absolutely gives to the problem of luck and happiness a mathematical image (necessarily very simplified), it is necessary to resort to that which I called the theory of dynamic probabilities.

One considers there two qualities which depend on one another and which depend also on chance; one of these qualities, that which the variation is the most rapid, represents luck, the other represents happiness.

These qualities are not united by a rigid link, they are simply connected; chance establishes between them a sort of elastic link; when one knows the value of one, one doesn't know the value of the other, but one knows the probabilities of all the values of the other.

When the one has a great value, the other probably has a great value.

This is precisely that which produces for luck and happiness, a simple connection uniting them, a very elastic link: a grand bit of luck probably renders a grand happiness, but does not render it for certain.

The assimilation is meanwhile very vague; if one acknowledges it, it will be happiness and not luck which will vary as the gains and losses of a game of pure chance.

As good sense indicates, the calculation is too precise and its scope too limited to be able to deliver a very useful contribution to a subject so ill-defined as luck and happiness.

Let us return to some considerations less abstract: it is ridiculous to attach a superstitious idea to the idea of good fortune and adversity or to believe that good luck and misfortune are, so to speak, inherited by an individual like physical or moral qualities.

Some seem pursued by adversity, and while this fact cannot be denied, it is not possible to conclude that a bad genie has attached itself to their kind; chance has been unfavorable toward them and their misfortune often is the cause of their unhappiness.

Certainly some common ideas will be very reasonable and very wise if, in the sense that one allows them, and doesn't confuse them with a superstitious thought.

One commonly says that a misfortune never arrives alone.

The idea of mystical nature that one attributes to this proverb is simply ridiculous, but the proverb, meanwhile, has an element of truth.

One misfortune often leads to another or renders it more probable; in the past, a war was always followed by famine or plague; a sickness could cause a great prejudice to some material interests. One could multiply these examples, the applicable proverb in such cases is incontestably sensible.

When it is a question of misfortunes having no dependence, the proverb, ridiculous *a priori*, is contradicted by the facts; it suffices from a little spirit of observation in order to convince oneself.

In general, one observes two successive misfortunes. When only one misfortune arrives, one doesn't note that it is not followed by another, so that the first case appears relatively more frequently than it is in reality. It is the kind of illusion on which Laplace expanded in his *Essai Philosophique* [*Philosophical Essay*]; this kind of illusion occurs everywhere; one notes the arrival of an event, one doesn't note its non-arrival.

Besides, those who are going to suffer an unhappiness find themselves in poor condition to observe, and if they attach a mystical idea to a proverb, the adverse proofs will change practically nothing. As it was well observed by Dr. Gustave Le Bon in his fine work on *les Opinions et les Croyances* [*Opinions and Beliefs*], the ideas of mystical origin, once anchored, are firmly established; the observation of facts doesn't have any more impact on them than having the most convincing and rational arguments.

— CHAPTER VIII —

MEAN VALUES

The notion of mean value has great importance, the mean value often gives, by itself alone, a general idea of all of the values.

We concede that the knowledge of a mean value cannot replace the knowledge of each of the elements that it summarizes, but that, for that same reason, a mean value may often be calculated independently from the knowledge of these elements.

In many of the cases one may determine a mean value without being obligated to solve the corresponding probability problem, as in many cases, one can get a general idea of a subject without being obliged to deepen the detail of it.

The notion of mean value is analogous to the notion of center of gravity in mechanics; as soon as one knows the motion of the center of gravity one can form an overall idea of the motion of a system; when one knows the mean value, one can form an overall idea of a result which ought to depend on chance.

Let us define in a precise and mathematical fashion that which one understands by the term mean value:

The mean value of a quantity is the sum of the products of different values of that quantity by the corresponding probabilities.

If a length is likely to take the three values 102, 106, 108, to which the corresponding probabilities are 1/8, 4/8, 3/8, the mean value of the length is $102 \times 1/8 + 106 \times 4/8 + 108 \times 3/8$ or 106.25.

One can say that the mean value of a quantity is the mathematical expectation of a player who should receive a sum equal to that quantity.

One sees, by the preceding example, that the mean value of a quantity is not generally a possible value of that quantity, just as the center of gravity of several material points isn't generally situated at one of those points.

One must not confuse the mean value of a quantity with the *most probable value* of that quantity (106 in the preceding example) nor with the *probable value* of that quantity.

The probable value is, by definition, as likely to be or not to be exceeded; one cannot always determine it exactly; it is thus in the preceding example.

(For around 20 years, some authors used at times the term of probable value instead of mean value. That manner of expression presented only some inconvenience, it seemed that one had invented it in order to put, at will, the definitions in contradiction with themselves and that one was contriving to maliciously introduce a confusion between the two terms of which the meaning is very clear. When after having defined probable value as mathematical expectation, one defined the mean difference and the probable difference, the mean error and the probable error, the mean life and the probable life, one used two contradictory definitions, or instead one gave to the same word two different meanings whereas it would be so simple to give to each word the unique meaning that fit it.)

(It is equally regrettable that one usually refers, in astronomy, to the expressions of error mean and mean error two quantities absolutely different. Here again it seems that one was contriving to use two terms lending to equivocation, then any confusion would not be possible in making use of the expressions of mean error and of squared error of which the sense define itself the same.)

The notions of mean value and mathematical expectation are analogous and likewise identical. Mathematical expectation is the mean value of a gain. Inversely, there is often advantage to assimilate the quantity of which one searches for the mean value of a sum of money, especially when one wants to appeal to the property of addition of the mathematical expectations.

The mathematical expectations which are some amounts of fictitious money are added as amounts of ordinary money.

The mean values are added equally, but their property of addition appears less evident.

One example, of which the result is incidentally very important, will show the interest that presents that property of addition.

The probability of an event is p at each trial, what is the mean value of the number of happenings of the event in m trials?

It is useless to calculate the mean value from its definition: let's imagine a player who is going to receive a franc when the event happened and who would receive nothing when it didn't happen; if the event happens n times, the player earns n francs, the mean value of the number of happenings of the event is the mean value of the gain of the player or the mathematical expectation.

The player has the probability p of earning 1 franc at each trial (or, if one prefers, at each round), his mathematical expectation for one round is p, and for m identical rounds it is mp.

The mean value of the number of happenings of the event is thus mp.

The contrary event, that is to say the non-happening of the considered event, has for probability $1 - p$, the mean value of the number of happenings of that contrary event is $m\,(1 - p)$, one could thus state this nearly evident law:

On average, the events produce themselves in numbers proportional to their probabilities.

We prove that that quantity mp is also the more probable value of the number of the happenings of the event of probability p in m trials (to the nearest unit, because mp is generally fractional); that quantity is thus, in some ways, the *normal* value of the number of happenings of the event.

In the preceding example, we have been able to calculate rather easily the mean value based on the definition; it will no longer be the same if the probability of the event varied from one trial to the other, if it was p_1 at the first trial, p_2 at the second,...p_m at the last. It would then be very laborious to directly calculate the mean value of the number of happenings of the event in the m trials.

On the other hand, if we imagine as previously a game which earned a franc when the event produces itself and which would earn nothing when it doesn't produce, the problem doesn't present more than the lesser difficulty; the sought after mean value is $p_1 + p_2 + \ldots + p_m$.

The problem, in some way the inverse of the previous, is equally interesting: if the probability of an event is p at each trial, the mean value of the number of trials that must attempted in order to observe the event producing itself is $1/p$.

If, for example, the probability of an event is 1/100, it would be necessary, on average, to attempt 100 trials for the event to produce itself. In all likelihood, a lesser number of trials will be necessary, because there is one chance in two that the event produces itself before 70 trials (in other terms, the probable value is 70), but the mean value considers cases where, therefore from the whim of randomness, it would take a very long time to observe the event producing itself.

We understand, for example, that it's necessary to keep from attributing to the expressions of mean value and of probable value a different sense from that which their definition confers to them: neither one nor the other corresponds in a necessary fashion to the vague idea of similar value.

One could make the same remark relative to the most probable value, it must be very far from all of the rest, it must be totally isolated, the same as it must be very slightly probable, albeit more probable than the other values.

§ 1. — PROBABLE LIFETIME

Let's imagine a group of 100,000 individuals of the age of 20 years; half of them will exceed the age of 66 years, the half of those who will survive more than 46 years. We say that the probable lifetime [life expectancy] at 20 years is 46 more years.

Let us consider a group of individuals aged thirty years; half of them will exceed the age of 68 years; the probable lifetime at 30 years is thus 38 more years.

In general terms, the "probable lifetime" of a group of a certain age is the number of years after which the group will be reduced by half.

That definition is in conformity with the general definition of the probable value; that is to say of the value which has an equal chance of being or not being exceeded.

§ 2. — AVERAGE AGE

Let us consider again a group of 100,000 individuals aged 20 years; some will survive one year only, some others two years, others three, etc. Let us call "survival" of an individual the number of years that he will live after his actual age of 20, or again, his total lifetime minus 20 years.

If one adds the survivals of these 100,000 individuals and if one divides the result by 100,000, in other words, if one takes the arithmetic average of survival, one obtains the "average life" at 20 years.

In general, the average life of a group of a certain age is the arithmetic average of survival of the individuals constituting the group.

One easily brings back that definition to the general definition of average values: the average life is the expected survival.

At 20 years, the average life is 43 years; at thirty years, it is 36 years.

At birth, the probable life is 55 years and the average life, 47 years.

(If, by the example of certain authors, one uses the expression of probable value instead of average value, one is obliged, after having defined the probable life as we have done, by specifying that the average life is the probable value of the life, that it is necessary to keep from confusing the probable life with the probable value of the life. I have already remarked, it seems that one contrives to introduce a confusion between some terms of which the distinction is very clear.)

— CHAPTER IX —

THE ORIGINS OF CALCULATING PROBABILITIES

Two very judicious questions, posed around two and a half centuries ago to Pascal by the chevalier de Méré, gentleman of spirit and famous player of games, had been the occasion of primary research on the calculation of probabilities.

One primary difficulty surprised M. de Méré: we have seen that the average value of a number of trials that are necessary to attempt to observe an event producing itself is the inverse of its probability. If for example, the probability of an event is 1/100, it's necessary to attempt, on average 100 trials in order for that event to produce itself, but in all likelihood the event will produce itself in a lesser number of trials; there is one chance in two that it produces before the seventieth trial and one chance in two that it produces after. The probable value of the number of trials is thus 70.

De Méré found paradoxically that the probable value was not proportional to the inverse of the probability. For example, when the probability is 1/2, there is equal chance for the event to produce itself at the first trial, or for it to produce itself later. After the idea of de Méré, one should be able to conclude that if the probability of the event is 1/100, there is an equal chance for that event to produce itself before 50 trials or for it to produce after, because 50 is to 100 as 1 is to 2.

The probable value is, as we have seen, 70 trials and not 50; it does not vary, as the average value, inversely to the probability.

The question posed by de Méré did not embarrass Pascal. Here is what he wrote to Fermat:

"I don't have time to send you the demonstration of a difficulty which truly surprised M. de Méré; because he has a fine mind, but he is not a geometer. It is, as you know, a big mistake. He told me thus that he had found difficulty in the numbers for that reason: if one attempts to make 6 with one die, there is an advantage to undertake it with four tries. If one attempts to 'ring the bell' (double sixes) with two dice, there is a disadvantage in attempting it in 24 throws, even though 24 is to 36, which is the number of faces of two dice, as 4 is to 6, which is the number of faces of one die.

"There is what was his grand scandal and which made him say proudly that the propositions were not consistent and that the Arithmetic was demented."

De Méré took as principle one idea which is correct when it is a question of the average value, but which is incorrect when it is a question of the probable value.

It is necessary meanwhile to recognize that the idea of de Méré, erroneous when it is a question of large probabilities, approaches more and more to the truth when one considers some smaller and smaller probabilities.

If the probability of an event is 1/100, the probable value of the number of trials that are necessary to make in order to see the event produce itself is 70, that is to say that one must wager equally that the event will produce itself in the first 70 trials or that it will only produce itself later.

If the probability is less than half, the probable value will be very nearly double; if the probability is less than a third, the probable value will be at least triple, etc.

In general, if the probability of an event is p at each trial, p was less than 1/100 (and practically the same at 1/25), the probable value of the number of trials that it is necessary to undertake to see the event produce itself is very nearly $.7/p$.

If, for example, the probability of an event is 1/400, one could wager evenly that that event will produce itself before 280 trials.

The average value would be in the same case 400 trials; one could again say that on average the event will produce itself in 400 trials.

§ 1. — Problem of Points

The second question posed to Pascal by M. de Méré has become classic under the name of "problem of the points." I am going to present this problem under a form which will make the delicate side understandable and will show that, following the famous words of Laplace, "one of the great advantages of the calculation of probabilities is to learn to defy the first impressions."

The players A and B play any game (to fix the idea, at cards). B is half as skillful as A, that is to say at each hand, A has two chances in three in winning, and B one chance in three.

In these conditions, it seems completely natural to think that A must render two points on four to his adversary; this is however inexact.

If he gives two points in four, he would have only 112 chances of winning against 131.

If he renders to him three points on six, he would have only 3.072 chances to win against 3.489.

If the number of hands becomes rather large, the player A would be able, without serious damage, to concede half of the points to his adversary; he cannot without disadvantage when the number of the hands is small.

Pascal resolved the problem of the points and proposed it to Fermat who resolved it from his perspective and went likewise even farther; Pascal and Fermat are considered as the founders of the calculation of probabilities. Earlier, Cardan, Galilée and maybe some others understood the chances in the simpler cases in the games of dice, but one cannot consider the few notions that they could have had as constituting the basis of a science.

A short time after Pascal, Huygens published the first treatise on the calculation of probabilities. It was a small book containing the solution to several problems related to games.

Jacques Bernoulli and de Montmort are the first who had made known the general formulas for the resolution of the problem of points; the formulas obtained by these two savants are based on some very different rationales and are apparently very dissimilar.

Here is approximately the primitive statement of the problem of points: two players, C and D, equally skillful (that is to say having at each hand an equal chance of winning or losing), each placing 10 francs on the game; the winner will receive 20 francs.

The winner will be the first who earns five points.

For some reason, the players are obliged to quit before the game is terminated, player C has, for example, scored three points, and player D two points. In what proportion should they equitably divide the 20 francs?

It is evident that player C, who has only two more points to score while player D has three more, must receive more than 10 francs.

This so-called points of player C is the part of the 20 francs that he must fairly receive, it is his mathematical expectation; in this particular case, the expectation is 13 fr. 75.

Let us present again the problem of points under another form: a candidate at an exam solves, on average, two questions out of three, so that the probability that he resolves a question is 2/3.

At the exam, one poses three questions, and the candidate, in order to pass, must solve two.

It should not be believed that the candidate has one chance in two to pass; that would be nearly true if, asking him 30 questions, it required that he know 20 of them; the small number of questions benefit him in a very strong proportion, he has 20 chances in 27 to pass.

If, posing six questions to the candidate, it required that he know four, he would again have nearly seven chances in ten to pass.

The unfortunate candidates in exams or competition never fail to attribute to chance the reason for their failure. It is necessary to admit that the part left to chance in nearly all exams or competitions removes any real seriousness from them; it is necessary to at least double the duration of the writing tests to be able, maybe, to obtain a nearly equitable ranking.

It is always necessary to believe that, under the pretext of amelioration, one does not look to replace in part the influence of chance, which is indifferent and impartial, for something systematically absurd. The public, used to error and illogic, even worries very little of the injustice, but randomness displeases it. When one wants to discredit an exam, one compares it to a lottery; the argument is decisive.

I will keep myself from expanding on the question of competitions; still less would I waste my time to take on the process of the ways of recruitment of the officials; I will content myself to remark that one competition could be triply censurable: by its score, by its program, by its process.

It is deplorable by its score when it must definitively exclude from a career some men of great value who, for multiple reasons that one conceives, cannot take part in that competition.

I don't have to criticize the programs, quite fortunately their study doesn't return in our setting; certain rules, meanwhile, bear their own criticism, since one proves the need to transform them every two or three years.

Even admitting that the purpose of competition is excellent and the rules irreproachable, nearly always the result depends much too much on chance for one to be able to take it seriously.

The public, who consider the competition as an ideal process, from the point of view of the selection to the point of view of the fairness, reveals itself as somewhat optimistic. These preceding remarks present evidence of the nature of these illusions.

— Chapter x —

Martingales and Lotteries

Of all the possible combinations, of all the systems that could inspire the imagination of players, it is the martingale that allows the obtaining of the greatest rewards. On the other hand, it is the martingale which also most rapidly leads the player to his loss.

One is aware of the conditions of the game: the player constantly risks the totality of that which he possesses, primitive bettings and earnings, he plays thus the largest game possible; he can stop after a certain number of rounds or, in theory, he can continue to play indefinitely.

The ruin of the martingale player is certain if he doesn't set in advance a maximum number of rounds of play; if p designates the probability that he wins a round, the probability that he will win n rounds running is pn. That probability tends toward zero as n increases.

If the game is disadvantageous at each round, the martingale is disadvantageous; it is even more disadvantageous than all other combinations of an equal number of rounds played.

If the game is fair, the martingale is evidently fair, the probability of ruin of the player tends toward certitude at the same time that the hoped for gain is believed by him to be indefinite.

If the game is advantageous at each round, the martingale is advantageous, it is even more advantageous than all other combinations of equal number of rounds played.

This result is in complete contradiction with the rules of prudence. A player possesses 1,000 francs and plays 10 francs per round at a game analogous to heads or tails but benefitting him a little will be nearly certain not to be ruined and likewise to win a fortune.

Another player, in the same conditions, plays his 1,000 francs in one hand and does the martingale will probably not go far: he will not have one chance in a million for which he plays more than ten hands.

The conclusion of the calculation is meanwhile unassailable; the martingale is much more advantageous, prudence gives reason to the first player, the calculations show favor to the second.

If, by a fabulous stroke of luck, the second player is not ruined after playing 142 rounds, he will have won a pot of gold a hundred billion times larger than the Sun; that maybe, with such hope, the near certainty of winning some miserable millions?

The conditions of a game could be advantageous and unreasonable. This last form, not having a precise sense, cannot be translated into a formula; the calculation makes known the probability of the possible results of the game, and this is the reason for this conclusion.

The pure martingale, a geometric progression that which we are going to study, is so fast, the chances of gain are so minimal, that most players prefer less ascending martingales, in arithmetic progression, for instance.

The player having risked 1,000 francs and is in line to win 1,000 will not play 2,000 francs at the following round, but only 1,500 francs. If he wins again he will play 2,000 francs, if he wins again he will play 2,500 francs only, etc. He will be able thus to build a sort of reserve which will permit him to fight longer against the opposing wealth or even win, if possible.

When the game is fair at each round, it is fair in its entirety; any combination, ascending or descending martingale, arithmetic or geometric progression, or whatever, cannot change anything. In these conditions, if the player doesn't fix for himself in advance a maximum number of rounds or a maximum gain, his ruin, in the long run, is certain.

It is more certain and much more rapid if the game is disadvantageous at each round, as is the ordinary case for the games practiced in rounds, in gaming houses and casinos.

When the game is advantageous at each round, every combination is advantageous. In general, if the player does not set in advance a maximum number of rounds his ruin is not certain and one can even hold as impossible if he only ever plays a small fraction of his capital.

For a given number of rounds, the game is, from the mathematical point of view, even more advantageous because it is more dangerous. There is in this result nothing paradoxical, on the contrary; when the player wants, so to speak, to abuse some advantage which the game reserves to him and while removing a too great mathematical expectation, the danger appears.

The martingale is the unique cause of the huge fortunes, one never knows of any other origin; that the martingale takes the industrial, commercial or financial form, it is always, in reality, from a game.

In order to become very rich, it is necessary to be favored by some competition of extraordinary circumstances and by some constantly favorable circumstances.

Never did a man become very rich by his merit.

§ 1. — THE LOTTERIES

Lotteries, in France, are organized, almost exclusively, by some work of beneficence, the ticket of 1 franc is ordinarily only worth a third of that amount. The buyer cares little about that; in paying for his ticket, the idea isn't coming to him of making a wise speculation, but more that of accomplishing a duty of charity: that he buys it, this is in reality a bit of hope, the possibility, if he has some imagination, to "build some castles in Spain," the certainty also of experiencing some emotion while reading the list of the drawings.

All things considered, for the buyer it is worth the money; a prize of one million draws more than ten prizes of 100,000 francs which would give him the same mathematical expectation but which wouldn't permit him the same dreams.

Before the Revolution, one lottery organized by the State under the name of Lottery of France functioned permanently. That lottery, suppressed in 1793, reestablished in 1797, was discontinued in 1839.

Several combinations were possible, the least disadvantageous, the *extrait simple* [*a single number drawn*] gave one chance in 18 of winning 15 times the wager.

The *extrait determiné* [*determined drawing*] gave one chance in 90 of winning 70 times the wager. The *ambe* [*a pair of numbers*] gave one chance in 400 of winning 270 times the wager. The *terne* [*three numbers*] gave one chance in around 12,000 of winning 5,500 times the wager and the *quaterne* [*four numbers*] one chance in around 500,000 of winning 75,000 times the wager.

One sees in such proportions that the lottery was disadvantageous for the public.

In the past certain States organized some lotteries and justified them by some reasons in the kind of the following of whose appreciation I leave to the reader: "There is nothing more desirable than the fortune of the people and the efforts of the government should aim towards the realization of this happiness. By what process is this achieved? The amelioration of material well-being is laborious and painful; there is a means of giving some happiness to the humble, it is to give them some hope. The lottery alone can give them the hope of becoming rich, of freeing themselves of all servitude and of realizing their ideal of liberty."

The lottery, according to these principles, was doubly useful, it procured at the same time a moral advantage to people and a pecuniary advantage to the State.

This point of view was shared by honest gentlemen, besides it was very tenable. The opinion which prevails today is that a government must not tolerate any lottery.

The State, in prohibiting these lotteries must, in good logic, oppose the issuing of bond premiums and likewise of securities with a high dividend premium. The State, in principle, should be opposed to gambling, wanting to suppress a utopia. Every commercial operation, financial or industrial entails hazards, it is always thus, from a certain point of view, comparable to a game.

The amounts wagered in racing and in the gaming houses are increasing year by year following a regular and rapid progression, more rapid maybe than the progression of the public fortune. It doesn't seem that the actual evolution has a very beneficial influence on the passion of the game, the complexity of modern life favoring worsening aggravation.

— CHAPTER XI —

CLASSIFICATION OF PROBABILITIES

One can divide the problems of calculation of the pure probabilities into the two following categories that, in enunciating them, include some exact data or some experiential data.

One tosses a coin at random twice in a row, what is the probability that both times it will show tails?

If we know for certain (this is the statement that must be said or implied) that the piece has an equal chance of falling on one side or another, the problem belongs in the first category; the probability in question is 1/4.

The problem will be in the second category if, ignoring the respective chances of the appearance of heads and of tails, the experiment we have conducted has supposed that these chances are equal.

If, for example, having tossed the coin ten times, heads shows five times, we would adopt, lacking any other information, the value 1/2 for the probability of the happening of heads, but this adoption would be made without prejudice and the nature of the problem would find itself modified; the probability of obtaining tails twice in a row would be greater than 1/4.

The problems of the first category could themselves be divided into the three following classes as:

The number of tests are not very large, the number of alternatives at each test is finite.

The number of tests is not very large, the number of alternatives at each test is infinite.

The number of tests is very large.

In the first case, it is an issue of discontinuous probability.

In the second case, it is an issue of semi-continuous probability.

In the third case, of a problem of continuous probability.

We have occupied ourselves exclusively until now with discontinuous probabilities. The study of semi-continuous probabilities isn't very important, I would say some words only of the principal questions of her traits: the problem of the composition of observation errors when the experiments are not very numerous and the problem of geometric probabilities.

The third case, that of very numerous tests, seems to get back into the two preceding; the number of tests was not very large at first, one could suppose in effect that it grows increasingly.

But first one difficulty usually presents itself; the complexity of the formulas becomes inextricable from that which is likewise from a very small number of tests, so that it will be impossible to draw any conclusions if the number of tests were large, which is precisely the more interesting case.

When the research question, owing to its extreme simplicity, permits the use of applicable formulas of a large number of tests, these formulas are practically incalculable and can't even give an overall picture on the solution of the problem.

If, on the other hand, one assumes *a priori* a large number of tests, this results in very favorable consequences: these large numbers tidy up some things, they soften the angles, they allow the neglect of all the vain details and conserve only the essential characteristics. All the analogous problems are merged into the same type, they differ only by their coefficients.

One conceives thus the generality of the law of large numbers; the simpler understanding, to it alone, nearly all the applications of calculation of probabilities.

When one considers a large number of tests, one equates that number of tests to a quantity able to vary in a continuous fashion.

While supposing, *a priori*, that continuous quantity of the sort that one could equate to time, one is conducting according to the *theory of continuous probability*.

One could say that the problems of probability divide themselves into three classes following whether they assume the discontinuity in time and space, the discontinuity in time and the continuity in space or finally the continuity in time and space.

This way of speaking is imaginative but incorrect, it is necessary to know how to interpret it; in reality, it is not usually a question of either time or space.

§ 1. — Geometric Probabilities

Buffon claimed to owe the purity of his style to the whiteness of his lace cuffs; it is without doubt the purity of his style which he primarily values as a writer; one generally denies to him the recognition of the qualities of a superior genius.

It is not certain here that we will be able to permit ourselves to express an opinion on the subject; it is not less certain that Buffon, in order to have been a naturalist and not a geometer, knew to be abstract at times. In the beginning, he resolved a problem of geometric probability.

"The analysis is, he says, the only instrument which one used until that day in science of probabilities to determine and fix the relationships with randomness: geometry appears not very tidy for a work so nimble; meanwhile, if one regards this closely, it will be easy to recognize that that advantage of the analysis of the geometry is entirely accidental, and that randomness, by virtue of it is modified and conditioned, finds resilient to the geometry as well as that of analysis...In order to put therefore the geometry in possession of these laws on the science of randomness, it is only a matter of inventing some games which roll on the entities and on the relationships."

The principal problem resolved by Buffon has become classic under the name of "problem of the aiguille [needle]"; here is the statement verbatim:

"I'm assuming that in a room of which the parquet flooring is simply divided by some parallel joints, one throws a baguette into the air, and that one of the players bets that the baguette will not intersect with any of the parallels of the parquet, and that the other player, to the contrary, bets that the baguette will intersect some one of the parallels; one asks for the fate of these two players (one can play this game on a checkerboard with a sewing needle or a headless pin)."

In designating by a the distance of the parallel joints and by r the length of the baguette (or the needle), assumed to be less than a, the probability for which the baguette meets one of the joints is given by

$2r/\pi a$, $\pi = 3.1416$ designating the ratio of the circumference to the diameter.

Generally speaking, one says that a problem is relative to the geometric probabilities when it consists to determine the probability for a set of points, of lines or of surfaces depending on a certain manner of randomness possessed by a given geometric property.

We have defined the probability by the relationship of a number of favorable cases to the total number of possible cases, all these cases having an equal likelihood.

It results from this definition that the first data of all problems of probability must be relative to the way of division in case of equal likelihood.

In most of the problems, this manner of division is implied in the statement; but when it is a matter of geometric probabilities, it is necessary, however, to specify it. Most modes of division can, in effect, appear acceptable after the statement of the problem, albeit generally they don't appear equally natural.

Relative to this subject, one example attributed to Bertrand has become classic: one traces randomly a cord in a circle, what is the probability that it will be smaller than the side of the equilateral triangle inscribed?

The expression of the randomly traced cord does not specify sufficiently the division in cases of equal likelihood; in adopting three modes of different division, Bertrand was led to three different values for the sought after probability: 1/2, 1/3, 1/4.

If one studies a particular question, the nature of the question can impose a mode of division; but if it is a matter of a general question, it is necessary to compare the different hypothesis which led to the different modes of division and to disengage from that comparison the hypothesis which seems the more natural, if there is one.

This comparison, which could be very delicate in certain cases, diverges somewhat from the domain of pure mathematics.

The problem of which I am going to say some words is the only one which presents a real interest from the geometric point of view and which permits discussion of diverse hypothesis.

We are going to see that, for the problem under consideration, one hypothesis seems totally natural. We will first deal with the other cases, before proceeding, in some way, by elimination.

It is a matter of studying the triangles traced at random, and at first defining that which one must understand by the expression of *triangles traced at random.*

Let us consider a flat area bounded by a closed curve F, let us decompose this area into a very large number of infinitesimally small squares by a double system of equidistant parallel lines; we can make at each square a corresponding ordered number; if we draw at random one of those numbers, as in a lottery, the infinitesimally small square, that is to say the corresponding point, is a point taken at random from the interior of the area.

If one draws three numbers and if one joins the corresponding points by some lines, one obtains a triangle.

This triangle depends, from a certain point of view, on chance, but cannot evidently be considered as a triangle traced at random.

While discarding the considered hypothesis, we must not conclude that it cannot present some interest or likewise be useful in certain cases; one of the more studied problems of the theories of geometric probabilities consists, precisely, to determine the average area of the triangle formed by three points taken at random in the interior of an area.

The relationship of that average area to the total area is 35/48 π2 = 0.07387 for the circle or ellipse, 11/144 = 0.0763 for the square or the rectangle, 1/12 = 0.0833 for the triangle, etc.

One can again obtain a triangle in breaking at random a stick into three pieces or, if one wants, by dividing the line AB into three segments by two points taken at random.

By this definition, it is necessary to understand that the line AB was divided into a very large number of equal elements, one made an ordered number correspond to each element. One drew at random two numbers, as in a lottery; the corresponding elements, that is to say the corresponding points, are some points taken at random on the line AB.

The three segments thus determined can only form a triangle if the largest is less than the sum of the two others. The probability for which the segments can form a triangle is 1/4.

In these conditions, the average triangle is that for which the sides are proportional to the numbers 16, 13, 7.

When the three segments form a triangle, the probability for which this triangle would be obtuse, is 9-12 log2, being about 0.682.

These results present some interest, but if the triangle formed by the breaking of a stick is due, from a certain point of view, to chance, it is evident that one cannot consider it like a triangle traced at random.

Instead of obtaining a triangle, in dividing a line AB into three segments by two points taken at random, one could divide three equal lines AB, A'B', A"B" into two segments while taking on each of them a point M, M', M" at random.

These segments AM, A'M', A"M" only determine a triangle if the largest is smaller than the sum of the two others; the probability of that eventuality is 1/2.

In these conditions, the average triangle is that whose sides are proportional to numbers 6, 5, 3.

The probability for which the triangle would be obtuse is (π - 2)/2, being about 0.571.

These results present some interest, but it is evident that the triangle obtained, as has been said, is not that which one could consider like a triangle traced at random.

So what do we mean by the expression by a triangle traced at random?

"A triangle traced at random is a triangle formed by three straight lines drawn at random, that is to say by three lines whose directions are absolutely anywhere."

It is on the direction of the sides that it seems natural to focus the ambiguity of randomness.

The material realization of the hypothesis would be considered by the throwing of three needles at random on a plane, as with the problem of Buffon.

The study of the question raised by the very interesting results for the geometry of the triangle: *the average triangle has for angles 110o, 50o and 20o.*

The probable value of the largest angle (that is to say the value which has as much chance of being or not being exceeded) is 106°. The probable values for the average angle and for the smallest angle are respectively 52° and 17°.

The probability that a triangle would be acute, that is to say that it had its three angles trebled, is 1/4.

When a triangle is acute, the average value of the largest angle is 80°, the value of the average angle is 65° and the average value of the smallest angle is 35°.

When the triangle is obtuse, the average values of the angles are 120°, 45° and 15°.

— Chapter XII —

Laws of Large Numbers

The laws of large numbers have very great importance, their applications extend to all the calculations of probabilities and one can almost say that they epitomize that which is truly essential in that calculation.

The laws which we will be studying here, the most simple and the most useful, require, to be well understood, but a few moments of attention and meanwhile, through their understanding, we will be capable of forming a general idea of many of the manifestations of randomness.

The laws of large numbers are not absolutely exact, in the most rigorous sense that one could attribute to that term, they are "asymptotic," that is to say that they approach more especially from the truth that the number of trials or rounds is very large; it is likewise for this reason that one calls them: laws of large numbers.

Practically, one could consider them as accurate, even when the number of trials is not very large.

In order to consider the simplest case, let's suppose that a player A should play a thousand rounds of heads or tails, the bet being 1 franc per round.

The player has an equal chance of winning or losing. This case, normal in a way, will be one in which he will finally have neither gain nor loss at the end of a thousand rounds.

If at the end of a thousand rounds, the player has gained 10 francs, one says that the difference is 10. If he has gained 24 francs, one says that the difference is 24. If he has lost 16 francs, one says that the difference is -16. If he has lost 28 francs, one says that the difference is -28, etc.

Since the player will constantly have an equal chance of winning or of losing, the probability of some positive difference is the same as the probability of the negative difference of the same amount. For example, the difference -8 has the same probability as the difference +8.

One conceives that the more a difference is large in absolute value, the more its probability is weak; I cannot demonstrate here this truth which incidentally is almost obvious.

Among all the differences, there is one that is particularly interesting, it is the *probable difference*; we name this difference thus, as it is equally likely of being or not being exceeded.

In the case being studied, the probable difference is ±21,

that is to say that there is one chance in four that the player loses more than 21 francs, one chance in four that the he loses less than 21 francs, one chance in four that he wins less than 21 francs and one chance in four that he wins more than 21 francs. In other words, the difference ±21 has an equal probability of being or not being exceeded.

Likewise, the difference ±30 has one chance in three of being exceeded, that is to say that the player has one chance in six of losing more than 30 francs, two chances in six of losing less than 30 francs, two chance in six of winning less than 30 francs, and one chance in six of winning more than 30 francs.

Likewise, the difference ±37 has one chance in four of being exceeded, that is to say that the player has one chance in eight of losing more than 37 francs, three chances in eight of losing less than 37 francs, three chances in eight of winning less than 37 francs and one chance in eight of winning more than 37 francs.

Regardless of the number of rounds, that which one calls the difference, for a fair game, is always simply the gain or the loss of the player; the difference is positive when it is a matter of a gain and negative when it is a matter of a loss.

We are going to see that the difference ±21 has one chance in two to be exceeded and that the difference ±30 has one chance in three of being exceeded. The ratio of the two differences is 30/21.

The law of large numbers, for this particular case, could be expressed thus:

Regardless of the game considered, provided that it is composed of equitable rounds, independent and numerous (not necessarily identical), the ratio of the difference which has one chance in three of being exceeded to the difference which has one chance in two of being exceeded is always 30/21.

The difference ± 37, in the chosen example, has one chance in four of being exceeded and the difference ±21 has one chance in two of being exceeded. For that particular case, the law of large numbers could be expressed thus:

If one must play a very large number of rounds (independent, but not necessarily identical) at any fair game, the ratio of the difference which has one chance in four of being exceeded to the difference which has one chance in two of being exceeded is always 37/21.

The law of large numbers is, so to speak, the key to calculating probabilities; some examples are very useful in order to make clear the meaning. We are going to first, without fear of monotony, consider the proportional differences at 21, 30 and 37.

One player has equal probability of winning or losing 5 francs at each hand; he must play 10,000 hands. The difference which has one chance in two to be exceeded (probable difference) is ± 340 francs, thus the difference which has one chance in three of being exceeded is ± 340 × 30/21 = ± 490. The difference that has one chance in four of being exceeded is ± 340 × 37/21 = ± 600.

A player, at each hand, has a probability of 1/3 of winning 12 francs, a probability of 1/3 of losing 9 francs and a probability of 1/3 of losing 3 francs; his game is fair. He must play 7,000 hands. The difference that has one chance in two of being exceeded is ± 158 francs, thus the difference that has one chance in three of being exceeded is ± 158 × 30/21 = ± 224. The difference that has one chance in four of being exceeded is ± 158 × 37/21 = ± 280.

One must play 3,000 hands. For 1,000 hands, one has one chance in two of winning or of losing 1 franc. For 1,000 others, one chance in two of winning or losing 3 francs. For 1,000 others, one has a probability of 1/4 of winning 8 francs, a probability of 1/4 of losing 2 francs and a probability of 1/2 of losing 3 francs; the game is fair.

The difference that has one chance in two of being exceeded is ± 186 francs (the order of the hands is not important), thus the difference that has one chance in three of being exceeded is ± 186 ×

30/21 = ± 266. The difference that has one chance in four of being exceeded is ± 186 × 37/21 = ± 328.

We understand by these examples, that which is, in general, the law of large numbers:

The ratio of one difference that has a certain chance of being exceeded by another difference which has another certain chance of being exceeded is always the same, regardless of the number of hands, provided that it is very large.

One doesn't assume that the hands are identical, one supposes only that they are equitable and independent.

We have chosen the differences proportional to 21, 30, 37, the same conclusions would evidently apply to other differences, for example:

The difference which has one chance in ten of being exceeded is always equal to the difference which has one chance in three of being exceeded multiplied by 3.77.

Regardless of the game considered and the number of hands, provided that it is very large, the ratio of the difference which has one chance of seven of being exceeded relative to the difference which has one chance in five of being exceeded is always equal to 1.17.

The ratios of the differences which have some given probabilities of being exceeded are always the same.

This is the law of large numbers, we can admire its extreme simplicity.

Knowing, for one game, the probability for a certain difference being exceeded, one can then deduce the probability for another given difference being exceeded.

One has constructed some tables making known the probability for a given difference being exceeded, that difference being expressed in taking the probable difference as unity. The probable difference, as we know, is that which has an equal chance of being or not being exceeded.

The probability for which the double difference of the probable difference being exceeded in one sense or in the other (in gain or in loss) is 0.177. The probability that the triple difference being exceeded is 0.043.

There is not one chance in ten millions that the difference will be superior to eight times the probable difference.

(For the mathematicians, I would say that the law of large numbers is not characterized by the constancy of ratios, but according to the analytical form of the expression of probability, form which is exponential; transcendent by consequence, but very simple.

Again for the mathematicians, I would say that the independence of the successive hands leads to the exponential law but that one could, in certain cases, be led to that same law without necessarily assuming independence.

When one doesn't make that assumption, one is led exceptionally to the exponential law and generally to some laws of very grand complexity.)

§ 1. — Generality of the Laws of Large Numbers

If thinking that the law of large numbers is relative only to games, one would be making a very inaccurate idea about their importance.

Every time that a quantity is susceptible, with equal chance, from augmentation or from diminution, the sum of these augmentations and of these diminutions follows, at the end of a large number of tests, the law precisely named the law of large numbers.

This is what happens, for example, when it is a matter of accidental errors of observation, one doesn't consider a single gain or loss but from other scales; the law of large numbers is applicable to them as we will see.

§ 2. — Uniform Tests

Let's consider the ordinary case where the successive rounds of a fair game are identical (to the close effects of chance). The theory leads to a general law which is infinitely simple, but which it is necessary to know how to interpret:

The differences are proportional to the square root of the number of rounds.

Let's take the first example, the case where one plays heads or tails, the wager was 1 franc per hand.

Player A has previously played 1,000 hands, let us imagine that he must play 4,000 of them.

The number of hands was four times as many, the differences, from the law of large numbers, would be two times more, since two is the square root of four.

Let's explain that which is necessary to understand this: the difference ±21 has one chance in two of being exceeded (probable error); in the case of 4,000 hands, the difference ± 2 × 21 or ± 42 will have one chance in two of being exceeded.

The difference ± 30 has one chance in three of being exceeded; in the case of 4,000 hands, the difference ± 2 × 30 or ± 60 will have one chance in three of being exceeded.

The difference ± 37 has one chance in four of being exceeded; in the case of 4,000 hands, the difference ± 2 × 37 or ± 74 will have one chance in four of being exceeded.

Let's suppose that the player must play 9,000 hands.

The number of hands is nine times larger than if he needed to play 1,000 hands, the difference, following the law of large numbers, will be three times greater than in the last case, since three is the square root of nine.

The difference ± 21 has one chance in two of being exceeded at the end of 1,000 hands, the difference ± 3 × 21 or ± 63 has one chance in two of being exceeded at the end of 9,000 hands.

The difference ± 30 has one chance in three of being exceeded at the end of 1,000 hands, the difference ± 3 × 30 or ± 90 has one chance in three of being exceeded at the end of 9,000 hands.

The difference ± 37 has one chance in four of being exceeded at the end of 1,000 hands, the difference ± 3 × 37 or ± 111 has one chance in four of being exceeded at the end of 9,000 hands.

These examples lead us to understand the sense that it is necessary to attribute to these terms: the differences are proportional to the square root of the number of hands.

The differences ± 21, ± 42, ± 63 which, in the preceding examples, have the same probability of being exceeded are *isoprobables*.

Likewise, the differences ± 30, ± 60, ± 90, which have the same chance of being exceeded, are isoprobable.

The differences ± 37, ± 74, ± 111 are likewise isoprobables.

To enunciate in a precise manner the law being studied, it would be necessary to say that the isoprobable differences are proportional to the square root of the number of hands. One ordinarily implies the word isoprobable to obtain a statement more concise and more elegant.

Not only the differences follow all the same law, resulting very fortunately, but that law is excessively simple, the differences are proportional to the square root of the number of hands; that is to say to the simplest function growing less quickly than the number itself.

If the number of hands increases more and more, the differences grow only proportionally as the square root of that number decreases indefinitely *relatively* to it, but they don't increase less in absolute value.

The ignorance of this last truth is the cause of one of the most frequent illusions of players.

I want to emphasize again on this fact, it is the isoprobable differences which follow the indicated law: it would be entirely incorrect to assume that the probability of winning 10 francs in 1,000 hands is equal to the probability of winning 20 francs in 4,000 hands. It is the probability of winning *more than* 10 francs in 1,000 hands that is equal to the probability of winning *more than* 20 francs in 4,000 hands.

The law of growth of isoprobable differences with the square root of the number of trials (the square root of time, when it is a matter of a continuous phenomenon) is, so to speak, the fundamental law of randomness; one finds it in studying speculation, errors of observation, kinematic and dynamic probabilities, Brownian motion,...one finds each time that it is a matter of a large number of independent and identical trials in which chance is alone and indifferent in one sense or another.

What must one understand by the expression *jouer gros jeu* [to play for high stakes]?

All the differences are proportional to the square root of the number of hands, so one plays even a bigger game if that quantity is bigger.

It is necessary to consider similarly a special game; the number of hands must be very large, a fair game is only identified by one characteristic: that is the square root of the mean value of the squares of the gains and losses for one hand. If, for example, at each hand, the player has the probability of 1/4 of losing 1 franc, the probability of 1/4 of losing 3 francs and the probability of 2/4 of winning 2 francs, the characteristic of his game is the square root of $1/4 \times 1^2 + 1/4 \times 3^2 + 2/4 \times 2^2$ or of 18/4, that is to say [$\sqrt{(18/4)}$, or…] 2.1213.

Two games can be different and assume the same characteristic; if one has one chance in two of winning 1 franc and one chance in two of losing 1 franc, at each hand, the characteristic is one.

If one has one chance in eight of winning 2 francs, one chance in eight of losing 2 francs and six chances in eight of breaking even, the characteristic is one.

The two games are very different if one considers a small number of hands; one could, without appreciable error, suppose them to be identical when the number of hands is very large.

Therefore the quantity which measures the fact of playing a more or less high stake game, or if one wants *the magnitude* of the game obtained by multiplying the characteristic coefficient by the square root of the number of hands.

If one multiplies the characteristic number of the game by the square root of the number of tests and by 0.68, one obtains *the probable difference*, that is to say the difference which has that much chance of being or not being exceeded.

Knowing the probable difference, one can calculate the probability for which a given difference can be exceeded, we have stated that. More generally, one can always calculate the probability to win a given sum at a game composed of a large number of independent hands, identical or not. That problem has been definitively resolved by Laplace.

§ 3. — Curve of Randomness

The number of hands being very large, one could, through the figment of imagination, assume that the differences are continuous.

The probability of a difference x (it is no longer a matter here of isoprobable differences) in a given number of hands is thus a continuous function of x, a function that I cannot study here. Its general character is one of decreasing with extreme rapidity when x becomes large enough; it is thus, as we have seen, that there is not one chance in ten million for which the difference would be greater by eight times the probable difference.

The representative curve of that function is called the curve of randomness or the curve of probability, it assumes the shape of a sort of bell more or less flared, following that the number of hands is more or less large.

On the figure [Fig. 1, below] are represented two curves of probability, one has plotted the differences on the abscissas and the corresponding probabilities on the ordinates. The curve ABC was relative to a certain number of hands, the curve A'B'C' is relative to a quadruple number of hands. The maximum ordinate, 0B, varies inversely to the square root of the number of hands.

The area contained between the curve and the x axis always has the value of one, since it expresses the sum of the probabilities of all the possible differences. If one assumes that the number of hands increases more and more, the curve continually flattens and finishes by nearly confusing itself with the x axis.

Regardless of the equitable game considered, the hands being identical or not, the curve by expanding always pass by the same shape; for mathematicians, one could say that the equation of the curve contains only one parameter, or again that all the curves belong to a same family.

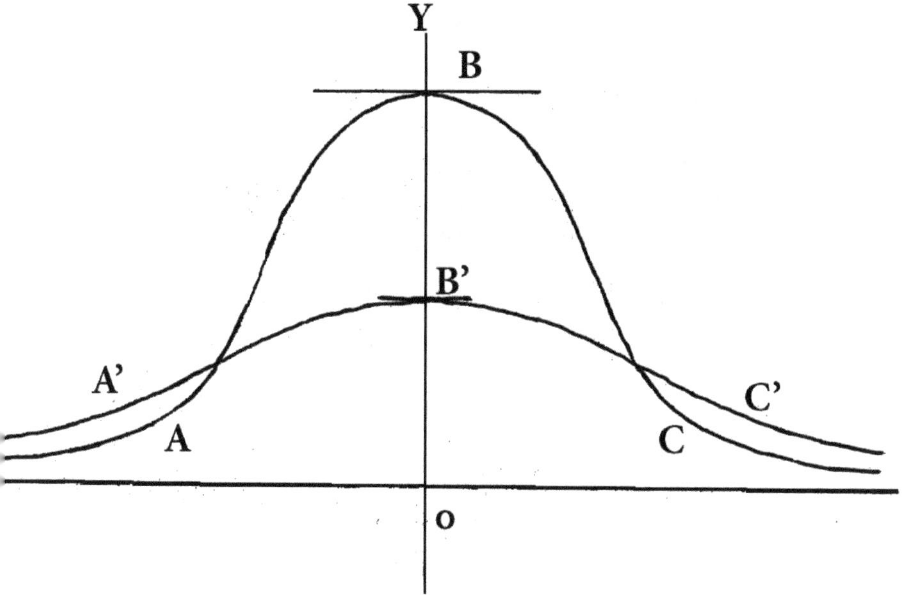

Figure 1

§ 4. — INEQUITABLE GAMES

When the game is equitable, the event, in some normal way, is that for which the player has neither a gain nor a loss. It is no longer the same when the game presents an advantage or a disadvantage, regardless how weak it may be.

Let's suppose that a player A must play a large number *m* identical hands of a disadvantaged game, we are going to thoroughly form an idea which may vary these gains or these losses.

We calculate at first the expected value or average value of gain of A for one hand, this quantity $-e$ is negative, since the game is disadvantageous.

The mathematical expectation for the *m* hands is the sum of the expectations for each of the hands considered in isolation, it has thus for a value $-me$, it is proportional to the number of hands.

The mean loss *me* is thus proportional to the number of hands.

That quantity *me* (due to the fact that *m* is a large number) is also the most probable value of the loss in *m* hands and also the probable value of the loss, that is to say the value which has as much chance of being or of not being exceeded.

The quantity *me*, mean probable value and most probable of the loss at the end of *m* hands is the value, sort of, *normal* for this loss.

It makes sense to compare all the possible losses or the gains to the normal loss; the difference between a given loss and the mean or normal loss carries the name of écart.

Let's assume, to fix the idea, that the number *m* of hands is 1,000, the mean loss *e*, for one hand, was 0 fr. 01; the mean loss for the 1,000 hands is 10 francs. If the player lost 30 francs, the écart is equal to -20 francs; if he gained 25 francs, the écart is equal to +35 francs; if he lost 2 francs, the écart is equal to +8 francs, etc.

The écarts on both sides of the mean loss follow exactly the same law as in the case of a fair game, the isoprobable écarts are proportional to the square root of the number of hands.

On could consider the normal loss as not being in any way due to chance; chance produces the écarts relative to that normal loss.

Let's review that which is essential:

The mean loss is proportional to the number of hands played; the écarts are more or less proportional to the square root of the same number.

The écarts thus diminish *relative* to the mean loss when the number of hands increase.

An example will not be useless, I chose it due to the manner that it corresponds closely to the game of *petits chevaux* [little horses] frequently played in the casinos.

Let's assume that, for each round, a player has one chance in ten of winning and nine chances in ten of losing.

He bets 1 franc each round and if he wins he collects 9 francs (his net gain is 8 francs, since his wager is 1 franc).

Let's first calculate the expected value for one round: the player has one chance in ten of winning 8 francs, his positive expectation is 8/10 or 0 fr. 80. Having, at the same time, nine chances in ten of losing 1 franc, his negative expectation is -9/10 or -0 fr. 90. The total expected value for one round is -0 fr. 10. The mean loss is thus 0 fr. 10 per round.

If one plays 500 rounds, the mean loss will be 50 francs, with a probable écart of 41 francs, more or less; that is to say that one would have one chance in four of winning, or of losing less than 9 francs; one chance in four of realizing a loss between 9 and 50 francs, inclusive, one chance in four of realizing a loss between 50 and 91 francs and one chance in four of losing more than 91 francs. The probability of winning would be 0.20.

The positive expectation will be equal to 7 francs; it is against that sum that the player would have to abandon his chances of gain.

The negative expectation is 57 francs. The probability of realizing a gain of some importance is very weak, the player has less than six chances in 100 of winning more than 50 francs.

Let's assume that one plays 5,000 rounds, the mean loss is 500 francs, with a probable écart of 128 francs, more or less; that is to say that there is one chance in four of losing less than 372 francs and one chance in four to lose more than 628 francs. One has only four chances in a thousand of winning.

If one plays 50,000 rounds, the mean loss will be 5,000 francs, with a probable écart of 407 francs, more or less, the probability of winning will be extremely small, one would only have five chances in a hundred of losing less than 4,000 francs.

§ 5. — Roulette

The game of *petit chevaux* is very disadvantageous, the game of roulette much less and meanwhile, in the long run, the player ought necessarily to lose.

In roulette, there are 37 numbers; 18 of them correspond to the color black, 18 others to the color red, the 37th number is zero.

If the player has wagered on black, he wins an amount equal to his wager if black is hit; he loses his wager if red is hit; he loses half of his wager if zero is hit.

The game would be equitable if the zero didn't exist. The small disadvantage resulting from the eventual outcome of the zero is of little importance when he played only a few rounds, but it becomes considerable in the long run.

The player has thus, at each round, the probability of 18/37 of winning 1 franc (or one whatever unit), the probability of 18/37 of losing a franc and the probability of 1/37 of losing 0 fr. 50.

If he plays 1,000 rounds, the normal loss is 13.50 and the probable écart is 21.50, more or less; the probability for which he would be ahead is 0.34.

If he plays 10,000 rounds, the normal loss is 135 and the probable écart is 68, more or less; the probability for which he would be ahead is 0.09.

In playing 40,000 rounds, the normal loss is 540 and the probable écart is 136, more or less; the probability of a gain is 0.0035.

§ 6. — Thirty and Forty

A great mathematician, Poisson, one of the creators of mathematical physics, had calculated the advantage of the banker at the game Thirty and Forty; the problem is unique, but interesting.

The game, somewhat more complicated, is analogous to that of Heads or Tails or of Roulette; it would be equitable if the banker did not have for himself the *re-do* for which the probability is around 0.022.

The re-do is analogous to the zero in roulette; when it occurs, the banker keeps half of the wagers of the players, his expected value is thus 0.011 for a wager equal to one.

The big shots can insure against the re-do by paying in advance 1% of his stake. Outside of this little premium, the game can be considered as fair and one sees that it is advantageous to insure, since the expected value of the players decreases thus from -0.011 to -0.010.

In playing 10,000 hands, with insurance, the player still has 16 chances in 100 to come out ahead.

In playing 40,000 hands, the probability of being ahead is .023.

§ 7. — GENERAL CONCLUSION

These examples demonstrate that the least disadvantage in a game renders impossible in the long run all chances of a net gain, it shows also that one can predict the loss with a smaller and smaller relative error.

As we have already noted, the normal loss is proportional to the number of hands and the écarts [errors] are, more or less, only proportional to the square root of the same number.

It is not always unreasonable to play, even if one doesn't have the temperament of a gambler; one can imagine some cases where one would prefer the possession of a 1,000 francs to one chance in two of possessing 2,000; the 2,000 francs could be indispensable for the intended purpose whereas 1,000 francs would be worthless.

In these cases, although exceptional, it is necessary to risk the thousand francs at once; in the playing in small increments, they are lost in advance.

In roulette, for example, one plays a single round nearly without disadvantage; in the long run, however, the loss is guaranteed.

Gamblers ordinarily only risk a minor portion of their purse, because they all have their system, their infallible combination in which they have special faith in that which is more complicated.

They have spent some treasures of the imagination in order to invent these systems, perhaps more like searching for perpetual motion and not less in vain.

All these infallible combinations, these ascending and descending martingales, these games in arithmetic and geometric progressions, all these are only naïve inventions for the mathematician who immediately perceives in them the weak points.

In reality, these marvelous systems infallibly lead the gambler to his loss; they lead there more or less rapidly, but in a manner always too certain.

In the long run, one can win at a game only if it is advantageous; if it is disadvantageous, one necessarily loses. Random chance, which is only capricious, cannot indefinitely wrestle against a constant cause which is slower, but of continual and certain manner.

§ 8. — Perturbing Effect of Chance

In thinking that our general conclusions are only applicable to games, one would have a distorted notion of the subject.

Every time that a constant cause intervenes at the same time as random chance one is led to the law of large numbers that we are going to study while assuming that it is a game. One could say that the constant cause produced an effect proportional to the number of trials, which is what one could call the normal effect. Chance produces a perturbing effect proportional to the square root of the number of trials.

That manner of experimenting is imprecise and, in all rigor, incorrect, but it gives a rather clear idea of the generality of the subject.

I will not study the case of a game that is neither fair nor uniform, that would lead us too far afield. I will not insist either on the principle of the *division of risks* put into practice by insurance companies. According to that principle, to the equal benefit of all, a company prefers a large number of small claims to a small number of large claims; one conceives, almost instinctively, that the role of chance is smaller in the first case than in the second due to the more numerous compensations.

— CHAPTER XIII —

BERNOULLI'S LAW

In the long run, the events produce themselves proportionally to their probability or nearly so.

Such is, reduced to the most popular form, the statement of the law of Bernoulli.

Let's be a little clearer. If the probability of an event is 1/3, that event will occur approximately 1,000 times in 3,000 trials. If it occurs 1,025 times, one says that the écart is 25. If it occurs 975 times, the écart is -25.

An écart equal to 25 in absolute value is hardly abnormal in this particular case; if one repeated the experiment, a greater écart could occur, on average, one time in three.

In 30,000 tests, the normal value of the number of occurrences of the event is 10,000. If the event occurs 10,250 times, one says that the écart is 250, if it occurs 9,750 times, one says that the écart is -250.

The number of tests is ten times larger than before. The écart, which has 250 for an absolute value and is ten times larger than the previously considered écart 25, isn't equivalent to it from the point of view of the probabilities; there are not three chances in a thousand for which the écart ± 250 will be exceeded in 30,000 trials.

In 300,000 trials, an écart of 2,500, proportional to the preceding écarts could be considered as impossible: there would not be one chance in hundreds of billions to be realized.

This example shows that the écarts on either side of the normal number of occurrences of an event do not grow proportionally to the number of trials, they certainly grow much less quickly.

Let's again clarify. Say that p, the probability of an occurrence at each trial, if one makes m trials, the mean value of the number of happenings of the event is mp. The value the most probable of this same number is again *mp*. That quantity *mp* is thus, in some ways, the *normal* value of the number of occurrences of the event.

If, in *m* trials, the event occurs in number $mp + x$, one says, for that reason, that the écart is *x* (that écart, consistent with these cases, is positive or negative).

If *m* is very large, that écart *x* is absolutely identifiable to one écart of gain or loss in a fair game.

All that has been said relative to the fair game applies in the studied case without modification.

The probable écart, that is to say that which has any chance of being or not being exceeded at the end of *m* trials is obtained by multiplying the number 0.68 by the geometric mean between the probability *p* of the event and the probability (1 - *p*) of the contrary event and finally in multiplying by the square root of the number of trials.

If, for example, the probability of an event is 1/3; in 3,000 trials, the normal value of the number of occurrences of the event is a thousand; the probable écart is: 0.68 × [(1/3)×(2/3)×3000]^(1/2)=17.6, that is to say that there is one chance in four for the event to occur less than 983 times and one chance in four for the event to occur more than 1,017 times.

We have seen that there was a probability of 0.043 for which the triple écart of the probable écart was exceeded, there were thus around two chances in 100 for the event to occur more than 1,053 times and around two chances in 100 that it would occur less than 947 times.

An event has for a probability 1/4. One must make 1,600 trials; the normal value of the number of occurrences of the event is 400.

The probable écart is: 0.68 × [(1/4)×(3/4)×1,600]^(1/2) or 12, there are thus one chance in four that the event occurs less than 388 times and one chance in four that it occurs more than 412 times.

We have seen that there was a 0.177 probability that the double écart of the probable écart was exceeded, there was thus around nine chances in a hundred that the event occurs less than 376 times and nine chances in a hundred that it occurs more than 424 times.

Not only the probable écart changes proportionally to the square root of the number of tests, but it is the same, more generally, for all the isoprobable écarts as we have seen whenever it was a matter of a game.

We could, while implying the word isoprobable, state the following law:

The écarts are proportional to the square root of the number of trials.

Let's suppose that the number of trials increases more and more, the écarts grow as the square root of the number of trials grow indefinitely in absolute value and decrease indefinitely in relative value.

This is, at lease in spirit, the law of Bernoulli.

One could again express it under another form. While supposing that the events occur in a proportional number to their probability, one commits an error. When the number of trials grows, the absolute error increases and the relative error diminishes.

§ 1. — LAW OF POISSON

That which preceded assumes that the probability of an event is the same at each trial. Let's assume now that it varies from one trial to another, but following a given law; it will have, for example, for value, p_1 at the first trial, p_2 at the second trial, ... p_m at the *m*th trial.

The mean value of the number of occurrences of the event in these m trials is, as we have seen, $p_1 + p_2 + \dots + p_m$. When the number m is very large, that mean value (except in some special cases), is at the same time the probable value and more probable; value, in some way, *normal* for the number of happenings of the event.

One calls the mean probability P the arithmetic mean of the probabilities, that is to say, the quantity $P = (1/m) \times (p_1 + p_2 + \dots + p_m)$

One can thus say that, in the long run, the event occurs in number proportional to its mean probability, or nearly so.

The normal value of the number of occurrences of the event in m trials is $p_1 + p_2 + \dots + p_m$ and if the event happens in number $p_1 + p_2 + \dots + p_m + x$ one says that the écart is x.

The écarts, in this particular case, follow the law of large numbers as the écarts in the equitable games, like the écarts in the case of the law of Bernoulli, where the trials are identical, but they are always smaller than they would be if one replaced the different probabilities $p_1, p_2 \dots p_m$ by their mean value P.

If, for example, one made 2,000 trials, if the event had for a probability 2/8 for the first thousand trials and 4/8 for the other thousand, the normal value of the number of occurrences of the event in 2,000 trials is 750 with a probable écart of 14.

If one makes 2,000 trials with the average probability 3/8 for the happening of the event, the normal value would be again 750, but the probable écart would be 15, it would have increased.

All the isoprobable écarts are, as one knows, proportional to the probable écart; thus, if one replaces some different probabilities by their average, the écarts are increased.

Let's imagine again that the successive values of the probabilities of the occurrences of the different trials are not exactly known. We know only that the probability depends on some constant causes and some periodic causes.

Thus, if the number of trials increase more and more and includes a great number of periods, the mean probability will approach more and more, by irregular oscillations but decreasing, to a fixed asymptotic value. All will nearly happen for a large number of trial as if the variable probability was replaced by its asymptotic value: the event will happen, on the whole, in a number proportional to that probability.

It is thus, it seems to me, that it is necessary to understand that which Poisson had named the law of large numbers; he had stated this law in a manner so vague and imprecise that no one, I believe, could ever understand exactly the idea that he wanted to express.

By the example of several authors, it appears useful to me to conserve rather happily the expression of the law of large numbers that Poisson had used but another meaning is attributed to it.

In reality, it is not always easy to recognize if the probability depends on constant causes (those which follow by the theorem of Bernoulli), or if these primary causes do not add to the other causes producing some periodic perturbations (theorem of Poisson), or if there doesn't exist some cause systematically making the probability vary in the same way.

§ 2. — Historic

The laws of large numbers are due to Jacques Bernoulli, de Moivre, Lagrange, Laplace, Poisson and Bienaymé. De Moivre and Laplace merit a special place.

For some years, some original conceptions and especially the idea of the movement of probabilities which gave an animated form, a sort of life, to the phenomenon of the transformation of probabilities, have given permission to the theory of large number to clear a new path. The phenomenon of pure randomness seems perfectly understood, but the related probabilities would uncover an indefinite horizon.

Since I mentioned some names, it is in all fairness to draft up a sort of list of masters allowing rendering justice to those who have most contributed to the progress of the science that we are studying.

The creators, the pioneers, Pascal, Fermat, Huygens and Halley, the illustrious astronomer to whom is credited the first mortality table, are beyond classification; it is necessary for us to admire them without trying to establish a parallel between their work and those of their successors.

Among the latter, the palm certainly goes to Laplace. His work, grandiose and sublime, likewise in the restricting of the subject which occupies us, has often been compared to those monuments which traverses the ages and which time respects.

After him it is necessary to cite Moivre, for whose profound work merits our admiration; it is to him who introduced into the calculus of probabilities the exponential formulas and who obtained the first results relative to the large numbers.

In third place, we could consider as being of equal merit: Jacques Bernoulli, de Montmort and Gauss.

And behind these grand masters, many others have contributed to the progress of calculation of probabilities: I cannot do the history here, as very succinct, from that calculation.

I would say only a few words of the principle works relative to our subject; the first serious book was the treatise by Huygens, of whom I have already spoken, it was entitled *De ratiociniis in ludo aleae* [*On Reasoning in Games of Chance*]. This little volume, published in 1657, studied some questions relative to games, it ends by describing some problems for which the author does not give the solution.

It was not until more than a half century later, in 1713, that the beautiful work of Jacques Bernoulli, *Ars conjectandi* [*The Art of Conjecturing*] appeared.

Five years previously [1708], de Montmort had published the first edition of his *Essay d'Analyse* [*Essay on the Analysis* (*of Games of Chance*)], but the text of Jacques Bernoulli had only been printed several years after the death of its author, and must be considered as having been written previous to the date of its publication. [Bernoulli 1713]

The Art of Conjecturing contains the resolution of the problems proposed by Huygens and the resolution of other questions on combinations and games. Bernoulli then demonstrated the celebrated theorem which keeps his name and on which he had, it appears, meditated for 20 years. Bernoulli ignored the asymptotic law of the increase of the écarts proportional to the square root of the number of trials; his objective was simply to prove that in supposing that the events occurred in proportional number to their probability, the committed error increases in absolute value and decreases in relative value when the number of trials grows more and more.

Presented under this model, Bernoulli's law appears nearly evident, but the demonstration of it is laborious and it greatly exaggerates its ingenuity.

In modern works, one no longer uses the demonstrations of Bernoulli, one contents oneself ordinarily with the asymptotic law: when the number of trials is very large, the écarts increase proportionally to the square root of that number.

That asymptotic law is not exactly that of Bernoulli, it assumes a large number of trials, that which the primitive law does not assume; on the other hand, it states an excessively general and simple truth that Bernoulli had always ignored.

One other demonstration only requires the knowledge of arithmetic; it is based on the additive property of the mean of squares. That demonstration does not exactly lead to the theory of Bernoulli, but to the following statement, which differs significantly: the mean value of the square of the écart is proportional to the number of trials.

That statement is correct; one can deduce from it the following conclusions, which are imprecise: the square of the écart is, "in total," proportional to the number of trials. The écart is, "in total," proportional to the square root of the number of trials.

We obtain thus, only with arithmetic, something analogous to the asymptotic law, but something very vague.

Some other demonstrations have been proposed, the examination of them will be fastidious; those which preceded are the most classic.

The work of Bernoulli was to end by the application to the moral and political sciences. What was the nature of these applications? We should always ignore them. Nicholas Bernoulli, who published the work after the death of his uncle, asked de Montmort to write a chapter on this subject. He excused himself; the more he meditated on the question, the more difficult it seemed to him to apply mathematics, which is precision itself, to a study rather vague as those of the moral and political sciences.

The first edition of the *Essay d'Analyse sur les jeux de hazard* [*Essay on the Analysis of Games of Chance*], by de Montmort, appeared in 1708. The work began with some philosophical considerations of randomness and probabilities; a difficult subject where one must avoid at the time of falling into the common and banal place or of rising up into overly pure abstraction. The ideas of de Montmort on randomness are those which one accepts nearly exclusively today; that is to say the non-existence of randomness as a creation of ignorant consciousness.

These generalities exposed, de Montmort studied the combinations and a large number of games now forgotten; already the general theory of the game is embryonic in his text as also in the work of Bernoulli.

The second edition, very much expanded from the first, appeared in 1713. Earlier, de Moivre had published a memoire entitled *De mensura sortis* [*On the Measurement of Chance*]; de Montmort (according to Fontenelle who wrote his academic eulogy) believed that this book was copied from his and was very piqued.

Thanks to the rivalry that thus rose up between the two scholars, the resolution of these problems that they asked themselves made rapid progress, also owing in large part to contributions of Nicholas Bernoulli.

The calculation of probabilities, which have hardly progressed since Huygens, overcame in five years one of the grandest stages; we can admire without reserve the ingenuity of these savants who had resolved, within two centuries, some problems that one considers still today as very tricky.

At the end of his book, de Montmort had the fortunate idea of reproducing his correspondence with Nicholas Bernoulli. This constituted a most curious document through the exchange of views which it contained on certain questions.

The *Doctrine of Chances*, by de Moivre, is one of the most beautiful works that has been published on the subject which occupies us; this book, like that of de Montmort, has not aged, although the third edition, the last, dates from 1756.

De Moivre, driven out of France by the revocation of the Edict of Nantes, fled to England, where he was soon considered a valued intellect. Newton gave him considerable support. The calculation of probabilities made for de Moivre the occasion of very profound research into pure mathematics, and the second place that we have accorded him in our recognition seems to have to come back to him in an indisputable way. The *Doctrine of Chances* contains some very beautiful developments on the general theory of games and the first calculations relative to the laws of large numbers.

We arrive at the grandiose *Théorie analytique des probabilités* [*Analytical Theory of Probabilities*], de Laplace, a masterful and transcendent work, unfortunately inaccessible even to most mathematicians.

The first edition appeared in 1812, it was soon followed by two others enriched by some additions and supplements. In 1842, the works of Laplace were published by the State. In the national edition, the analytic theory is preceded by the *Philosophical Essay on Probabilities*, from which I previously borrowed some material.

The philosophical essay is a beautiful work of severe elegance, in a simple but majestic style; in creating it, Laplace had the idea of popularizing the elements of the calculation of probabilities and in making the appreciation of the principles like the applications without resorting to the lesser mathematical formulas.

From the philosophical point of view, Laplace had attained the objective that he proposed and his *Essay* is well worthy of the success that it obtained, but, from the point of view of the scientific popularization, it has permitted the offering of some doubts on the value of his work and to assume that he had been some small victim of an illusion in thinking that a subject of mathematical nature could be exposed through ordinary language, without the aid of numerical examples.

It is necessary besides to recognize that the calculation of probabilities is a very difficult popularization; it is abstract in itself for the reasons that I have already explained and some simple examples are often necessary to comprehend the elements in them.

In the *Analytical Theory*, the style is concise, the analysis profound, but Laplace never seems to worry about knowing if the reader can follow it, the important results are not put sufficiently in evidence and the author, by his own admission, treats only the most difficult questions.

The work begins with two or three hundred pages of a stringent and abstract analysis before even giving the definition of probability. That way of proceeding, which consists of dealing first with the questions of mathematical analysis which could be useful later, so to speak, of "clearing the way," is to some advantage, but it doesn't ordinarily encourage the reader and often frightens him.

The *Analytical Theory* does not study exclusively the speculative portion of the subject, it treats also some applications, but while always keeping itself at a very high level.

One has sometimes criticized in the works of Laplace some questions of detail, some light errors; any human work lacks perfection. On the whole, the works of Laplace are worthy of admiration.

Some very important works have often been published under the form of memoirs; the enumeration of them would be long and fastidious, it is only necessary that we know that some of the books have conserved a great value from the historic point of view.

— Chapter XIV —

The Gambler's Ruin

Ever since the beginning of the analysis of chance, the problem of the gambler's ruin has been considered as presenting a superior interest from the mathematical point of view.

Its difficulty has not been a foreigner to its success, and among the most famous names who honor the calculation of chances, one can cite only Gauss who didn't bring any contribution to the study of a subject so captivating.

It is through a memoir on the ruin of gamblers that Ampère, the illustrious physician, revealed to the scientific world; the problem of the gambler's ruin is not that which one would think as common, it is a most beautiful question from the purely scientific point of view.

The new conceptions of the design of probabilities show well the interest of the problem; the probability is assimilated there as a sort of fluid of which the laws of propagation are more or less complex. After having studied the problem of the propagation in an unlimited space, it is totally natural to study it in the case of a finite space, this is the problem of the ruin of the idealized gambler.

The general theory of the game does not study any game in particular; it assumes that the conditions of the game are continuous for each round and it studies the sort of players at the end of a certain number of rounds.

Let us confine ourselves to the case of two players; the simplest problem that we have resolved, when it is a matter of a large number of hands assume that the players possess an infinite sum or at least equal to the sum that they would be able to lose if all the hands were unfavorable for them.

We are going to approach the study of a problem more difficult: the sum total that one of the players is going to risk at the game and that, in order to simplify, we will say his *fortune* is limited; the fortune of his adversary is assumed to be infinite or at least equal to the sum total that he would be able to lose if chance was constantly against him.

The hypothesis of an adversary of infinite fortune is feasible because, to play against every adversary who presents himself is, in reality, to play against an adversary who has an infinite fortune.

We would say, in order to simplify, that the player is ruined when he has lost the total sum that he dedicated to the game.

Let's assume that after each hand the loser pays his wager to the winner. Under these conditions, the player for whom the fortune is limited would be able, at a given moment, to have lost without hope of recovering the total sum that he wanted to play, he would thus be ruined.

We will consistently study the fate of the player who has a finite fortune; the fate of his adversary, of infinite fortune, is deduced besides immediately.

One could ask oneself about the probabilities of ruin of the player two kinds of problems of a degree of difficulty very different: one could assume that no limit is fixed for the duration of the game, it lasts until the ruin of the player or it lasts indefinitely if the player is never ruined. This first problem is very elementary.

The second problem consists of determining the probability for which the player is ruined before a number of given rounds. For example, the player A plays heads or tails for 1 franc per round and has 100 francs, we ask the probability for which he would be ruined before having played 10,000 rounds. This second problem, of which the first is only a particular case is very interesting but difficult.

Cases where no limit is fixed for the duration of the game. We have seen, in studying these applications of the notion of mathematical expectation, that the ruin of the player is certain if the game is equitable.

The example that has been given removes from that result any paradox. If one assumes that the fortune of the adversary of the considered player A was limited first and then she added indefinitely, the expected sum by that player A believes to infinity and, in all fairness, the probability of her ruin must reach toward certainty.

Thus, whenever a player plays equitably against any adversary who presents him/herself where, if one wishes, against the public who could be likened to a unique excessively rich player, her ruin, in the long run, is certain.

It will be certain even more so and much more rapidly if the conditions of the game are unfavorable to her.

It is no longer the same when the conditions of the game hold an advantage to the player, even if that advantage is weak.

Not only in this case is the probability of ruin no longer a certainty but, it is likewise very small if the player bets on each round only a small fraction of his capital.

We know the gross profit that are realized by the casinos on roulette and Thirty and Forty, which nonetheless are nearly equitable games. The probability of ruin, for these casinos, is practically nil; we are going to observe this.

If one defines as p the probability of ruin from a player A playing 1 franc per round at an advantageous game when he has only 1 franc, the probability of his ruin when he possesses a francs is pa.

The probability of ruin pa as long as the player possess a francs is very weak as soon as a is slightly large.

If, for example, the player has 51 chances out of 100 of winning and 49 chances out of 100 of losing at each round, the game doesn't seem hardly advantageous. If he possesses 100 francs, the probability of his ruin is only 0.018; if he possesses 500 francs, the probability of ruin is not more than two billionths.

That example shows how the least advantage changes the conditions of success of a game.

If the player possesses a francs and if he plays at heads or tails, the probable duration of the game is around $2.2a2$ (except when $a = 1$), that is to say that there are as many chances for which the game ends before $2.2a2$ rounds as chances for which the game continues for a long time. If the player possesses 10 francs, the probable duration is 220 rounds.

There are around 99 chances in 100 for which the player is ruined before having played 67 $a2$ rounds.

Cases where the duration of the game is limited. — The problem consists of finding the probability for which player A, who possesses the fortune a, is ruined before having played n rounds.

When the game is equitable, a theory of extreme simplicity permits the calculation of the probability of ruin:

The probability for which player A who possesses the fortune *a* is ruined before having played *n* rounds is twice the probability that it would be for losing a sum greater than *a* at the end of *n* rounds if his fortune was unlimited.

We know how to calculate the probability for which the loss *a* is exceeded at the end of *n* rounds, just double the figure to obtain the probability of ruin.

The player A possesses 21 francs and plays at heads or tails at 1 franc per round; what is the probability for which he is ruined before the thousandth round?

If he possesses an unlimited sum (or at least equal to a 1,000 francs), the probability for which, at the end of a thousand rounds, he loses more than 21 francs will be 1/4. The probability for which he is ruined before a thousand rounds is thus 2 × 1/4 = 1/2.

The player A possesses one 100 francs; he plays at heads or tails at 1 franc per round, what is the probability for which he is ruined before 10,000 rounds.

If he possesses an unlimited sum, the probability for which he loses more than 100 francs at the end of 10,000 rounds will be 0.155. The probability that he is ruined before 10,000 rounds is thus 2 × 0.155 = 0.31.

The theory is true not only when it is a matter of a game constantly identical to itself but again when the bets can vary at each round, the game remains equitable.

The case of an unequitable game is much more difficult: in order to resolve the problem of the ruin, it is necessary to have access to the formulas that I made known in my treatise; I will content myself to cite an example.

The player A possesses 100 francs; he plays a roulette at 1 franc per round, what is the probability that he is ruined before 10,000 rounds?

At roulette the player has 18 chances in 37 to win a franc, 18 chances in 37 to lose a franc and one chance in 37 of losing 0 fr. 50. The probability of his ruin in 10,000 rounds is 0.78.

The small disadvantage that presents the game (one chance in 37 of losing 0 fr. 50) has greatly increased the probability of ruin; if this small disadvantage didn't exist, the probability of ruin in 10,000 rounds would be 0.31.

If no limit is fixed for the duration of the game the ruin is certain, whether the game is disadvantaged or equitable, but it is much more rapid in the first case.

§ 1. — Maximum Écart

Attached to the theory of gambler's ruin is the interesting but difficult problem of the maximum écart, a problem that I had resolved some years ago.

Player A must play a great number of rounds at an equitable game; let's assume, in order to set the idea, that he must play a thousand rounds of heads or tails, it is the case of our first example relative to the law of large numbers. We have seen that the probable écart is ± 21 francs, that is to say that the player has one chance in four of winning more than 21 francs and one chance in four of losing more than 21 francs at the end of a thousand rounds; that is in other words, the écart ± 21 has an equal chance of being or of not being exceeded at the end of a thousand rounds. We know also that the écart ± 30 has one chance in three of being exceeded and the écart ± 37 has one chance in four of being exceeded at the end of a thousand rounds. We know equally that there is a probability of 0.177 for which the écart which is twice the probable écart is exceeded at the end of a thousand rounds and that there is a probability of 0.043 for which the écart which is triple the probable écart is exceeded at the end of thousand rounds.

The écarts that we have studied and which are regulated by the classic laws express some large numbers and are the écarts which exist *at the end* of a certain number of rounds.

The theory of the maximum écart studies the probabilities relative to the much larger écart which produces *in the course* of the thousand rounds, or more generally in the course of a given number (very large) rounds.

It is very difficult to know the probability for which an écart of 8 francs occurs at the end of a thousand rounds or the probability for which the largest écart which occurs in the course of a thousand rounds will be 8 francs. Very rarely the largest écart is realized at the last round; if the player has finally earned 8 francs, very likely he had gained more previously, very likely also, at a certain instant, in the course of the game, he had lost more than 8 francs.

The probability for which an écart is exceeded *in the course* of a certain number of rounds is necessarily larger than the probability for which that écart is exceeded *at the end of* that same number of rounds.

In the case of a thousand rounds played at heads or tails, there is one chance in two for which the largest écart which occurs in the course of the game is more than ± 36 francs and one chance in two for which it is less than ± 36 francs.

One says thus that the second probable écart is ±36 francs.

Generally speaking, one calls the *second probable écart* the écart which has an equal chance of being or not being exceeded *in the course* of the rounds.

We understand the difference between the two probable écarts; the first has equal chances of being or of not being exceeded at the end of n rounds which compose the game, whereas the second has equal probability of being or not being exceeded in the course of n rounds.

Let's express a first very interesting theory:

The second probable écart is equal to the first probable écart multiplied by 1.7.

This result assumes nothing about whether the rounds which compose the game are identical, it assumes only that these rounds are equitable, independent and numerous.

We have seen, in studying the classic laws of large numbers, that the écarts which have some given probability of being exceeded at the end of a certain number of rounds have some constant relationship, that being this number of rounds. Thus, the écart which has the probability of 1/4 of being exceeded (the probable écart) as 37 is to 21, regardless of the equitable game considered and the number of rounds of which it is composed, providing that this number is large.

It is very interesting to know if an analogous law exists for the écarts which occur in the course of the rounds.

For example, does the écart which has one chance in four of being exceeded in the course of the rounds have a constant relationship with the écart which has a one chance in two of being exceeded (second probable écart) in the course of the rounds?

The relationship is constant; in effect, it is the same regardless of the number of rounds provided that this number is large but is not equal to 37/21.

The consistency of the relationships exists as for the ordinary écarts, but these relationships are not the same.

By a stretch of the imagination for which we have already had recourse, one could assume that the écarts are continuous.

The difference between the probability for which the écart $x + dx$ is exceeded in the course of the rounds and the probability for which the écart x is exceeded in the course of the same rounds, is the probability for which the *maximum écart* that occurs in the course of the rounds is x.

The probability for which the maximum écart is $\pm x$ is a function of x that one can represent by a curve called the second curve of probability, it takes the form indicated in the figure [Fig. 2, below].

When the number of rounds increases more and more, the curve flattens while at the same time its peak moves to the right.

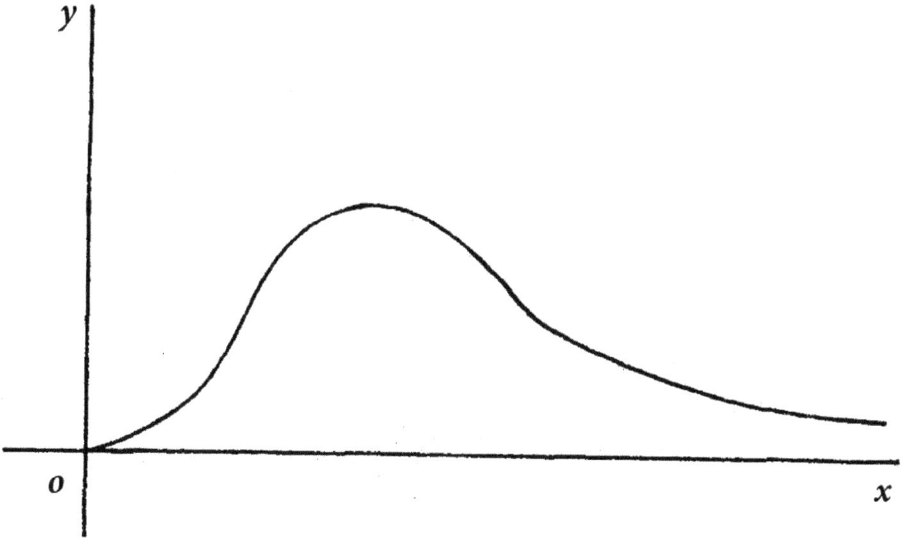

Figure 2

It is interesting to ask what the most probable value is of the largest écart which occurs in the course of the rounds.

If, for example, one has been promised to receive a certain sum, if one just guesses what will be the value of that écart, what value must one choose?

The most probable value of the maximum écart is equal to the probable écart multiplied by 1.34.

If one has been promised to receive a sum equal to the maximum écart, that expectation would have for a value the product of the probable écart times 1.94. In other words, the average value of the maximum écart is equal to the product of the probable écart times 1.94.

To summarize, if one takes as unity the probable écart, the maximum écart has 1.34 for a most probable value, 1.7 for the probable value and 1.94 for the average value.

Uniform game. — The preceding assumes only that the rounds which compose the game are equitable, independent and numerous. The case where the game is uniform, that is to say where the rounds are identical before chance differentiates them, is especially interesting.

We have seen, in studying the classic laws of large numbers, that the isoprobable écarts increase proportionally to the square root of the number of rounds. For example, the probability for which an écart x is exceeded at the end of 4,000 rounds and has the probability for which the écart $3x$ is exceeded at the end of 9,000 rounds.

One analogous law exists when it is a matter of some écarts which occur in the course of the considered rounds.

The probability for which the écart x is exceeded in the course of 1,000 rounds is equal to the probability for which the écart $2x$ is exceeded in the course of 4,000 rounds and at the probability for which the écart $3x$ is exceeded in the course of 9,000 rounds.

The écarts which have the given probabilities of being exceeded in the course of the rounds increasing proportionally to the square root of the number of these rounds.

— CHAPTER XV —

THE ILLUSIONS OF GAMBLERS

"It is particularly in games that the weakness of the human spirit is revealed and the inclination that the spirit has to superstition. Nothing is as ordinary as seeing some gamblers attribute their misfortune to the people who hang over them, and to the others circumstances which are no less indifferent to the outcome of the game. There are those who make a rule of taking only the cards which gain in the thought that a certain good fortune is attached to them. Some others, on the other hand, keep for themselves the loosing cards, with the opinion that having several times lost, it is less likely that they will lose again, as if the past could decide something about the future. There are those who invoke certain places and certain days. One sees in them who refuse to shuffle the cards if certain situations do not exist and who believe their loss to be inevitable if they hold those cards contrary to their own rules. Finally most of them look for their advantages where they don't exist, or even ignore them entirely.

One could say approximately the same thing about the conduct of men in all the actions of life where chance has some part. These are the same prejudices as the government; it is the imagination which regulates their advances and which makes blind at birth their beliefs and their expectations" (De Montmort — *Essay d'analyse sur les jeux de hazard*, 1708) [*Essay on the Analysis of the Games of Chance*].

In order to use the very clear classification of Dr. Gustave Le Bon, revealed in his beautiful work on *Les Opinions et les Croyances* [*Opinions and Beliefs*], one could say that the illusions of gamblers are of two types: some are of mystical nature, the others are of falsely rational nature.

The first, interesting from the psychological point of view, shows, as Montmort said it, the weakness of the human character.

The second proves that nearly always those players do not realize the *absolute independence* of the successive rounds.

As soon as heads is shown ten times in a row, they figure that at the following toss, tails has more of a chance to show itself.

As soon as red presents itself several times in a row at roulette, they consider the happening of black as nearly certain at the following round.

From where does this illusion come? It is very unlikely *a priori* that heads will show itself 11 times in a row and it is this improbability that suggests, when it has happened ten times, that it will not happen at the following round. The illusion holds mostly to that which one refers involuntarily to as the origin of the events when it isn't too remote. If, in the evening, the game was terminated by a series of ten reds, the players would have the idea of beginning by betting on black, the origin being too far away.

That question of the origin and of the relationship between the successive rounds leads the players to all kinds of errors. There is no origin, there is no relationship, the rounds are absolutely independent and have no bearing on one another.

Besides, we will see further on by some statistics made on roulette that the theory, which assumes the independence of rounds, is in perfect accord with experience.

Players generally figure, when a large écart happens in one direction, that the écart is going to have a tendency to diminish, thus obeying a sort of principle of compensation which is only an illusion. The rounds are independent, the past écarts have no influence at all on the future écarts.

We could translate into a formula the kind of game which would conform to that illusion of the players. I have already spoken of the theory of connected probabilities; in that theory one studies, among others, a game for which there is a tendency toward the reduction of the écarts. One arbitrary function of the time that contains the formula and which one could not make vary at will permits the realization of the degree of illusion of the player considered and more or less by the persistence of that illusion, by the variation with the time.

The law of Bernoulli, when it is not expressed under a form sufficiently explicit, does not guarantee the player against these illusions. On the contrary, it seems to reinforce by a rational argument, by a mathematical proof, an idea entirely inaccurate.

In saying that the events happen, in the long run, in a number proportional to their probability, one states a theory which needs, as we have seen, to be severely precise; the erroneous interpretation leads to misleading results and some dangerous errors.

It is necessary to fully understand that while considering an event before it happens, in a series of trials, in a number proportional to its probability, one commits an error. The more the number of trials is considerable, the more the error is small and the value relative and the more it is large in absolute value.

Very often the player is smitten by the first fact and loses sight of the second, that one which meanwhile should be the more interesting.

Regardless of the game considered, the total mathematical expectation is equal to the sum of the mathematical expectation of all the rounds considered in isolation.

The games ordinarily practiced, roulette, Thirty and Forty, little horses, etc., are disadvantageous for the player. The total disadvantage (total mathematical expectation) is equal to the sum of the disadvantages of each round considered in isolation.

All the possible combinations, all the imaginable martingales change nothing.

To think of it as a disadvantageous game at each round but which can be advantageous in aggregate is an error which bothers the mathematician and which seems to him inconceivable.

When a player is persuaded to have found a combination assuring a certain benefit, it is only by receipt of conscience that one must try to undeceive. The most probing reasoning does not shake his conviction and all the formulas of algebra won't alter his faith.

It is curious that the failure of the system and the loss that it causes does not discourage the true player; after the unsuccessful, he always thinks that one other combination, the real one, will permit him to regain the lost sum and to achieve at last the fortune. From one unrealized hope always follows another hope.

— CHAPTER XVI —

NEW THEORIES OF PROBABILITY

§ 1.— INFLUENCE OF PROBABILITIES

The probability which is an abstraction shines like a little sun.

The probability which is an abstraction disintegrates like a lump of sugar which we place in grog.

The probability born from an instantaneous source; it spurts out spontaneously, with infinite speed, since it diffuses slowly into space.

The probability is a sort of fluid forming from the waves which form and which stretch out like the swells of the ocean.

The probability is not only the relationship of two numbers for which the first is always smaller than the second; this is a sort of matter, of energy … of things which transform and which is, so to speak, animated and moving.

These analogies, these images, likewise vague and indistinct, present a very real interest which they don't hold only from their originality or from their contingency. Speaking to the imagination, they often permit better penetration of a subject too abstract; sometimes likewise, it permits a glimpse of new truths.

That a reconciliation could be unexpected and strange does not diminish the prize. Certainly to the contrary, the contrast in the similarity (if one could say) is, to use an analogy, an element of value which reinforces the interest.

In our world of probabilities, we must, as elsewhere without doubt, capitalize on all the advantages that allow us to come from behind.

This domain has the reputation of being very dry; that is the abstraction in all its dryness and in its cold rigor. It is necessary therefore to endeavor to acquire some new conceptions permitting us to better know the resources and the scope.

The theory of the radiance presents nothing of great difficulty; I cannot meanwhile, in this little book, expose the principles. In order to understand all the interest, it is necessary therefore to be acquainted with the theory of the transfer of heat. It is the link between those two studies which is especially curious.

§ 2. — Theory of Continuous Probability

The most important problems of the calculation of probabilities are evidently those which are related to some large number of tests.

The laws of chance cannot appear in their absolutely general form if the number of tests is rather large, in order that it produces a sort of fusion which will eliminate the many particulars in each question and to let subsist only the essential characteristics.

We have seen the extreme simplicity and the real beauty of the laws of large numbers; they assume, precisely, that this sort of fusion was effected.

In order to obtain the laws of large numbers, one assumes at first that it is a question of any number of trials, and then that number increases more and more, which permits some simplifications by using the famous formula of Sterling.

The theory of continuous probabilities proceeds otherwise: it does not go from small numbers to large numbers; it studies directly the case of large numbers, and likewise it replaces *a priori* these large numbers by a continuous variable that one could call *time*.

The introduction (optional elsewhere) of the notion of time permits us to conceive of the transformation of the probabilities as a phenomenon. The advantage is evident.

The theory of radiation is a particular case of the theory of continuous probabilities; one assumes there that it is in reality a matter of time and that the transformation of the probabilities often operated according to a very simple law. Ordinarily, the law of the movement of probabilities is more complex than those of radiation in particular.

One other advantage of the continuous probabilities is to permit, for all questions, the direct and immediate use of infinitesimal calculus. This calculus, much simpler than the calculus of finite quantities, permits the solving of problems which, by their complexity, would be unapproachable with ordinary methods.

The first merit of the theory consists in the number of new results which it has allowed to be obtained and in the degree of difficulty of the questions that it has addressed.

Outside of the supposition of the continuity, a second fundamental notion is that of the unity of the calculation of probabilities: all the relative problems of large numbers are taken back to a unique type and presented under the same form.

The interest of that unity of notion is evident: the different problems are presented under the same form, it is possible to compare them and to understand their true characters.

One could thus establish a natural and logical classification of the different problems; this classification, based on some real characteristics, use the point of view of studies as well as the point of view of the research, made from the theory of continuous probability a methodic and rational science.

One conceives that the reunion of these advantages has permitted the realization of much progress. My *Traité sur le calcul des probabilités* [*Treatise on the Calculation of Probabilities*] contains the systematic development of the theory of continuous probability; I cannot likewise try to give it a survey here, I can only summarize the two principles which constitute the basis: the supposition of the continuity of all the variables and the reduction of all the questions to a unique type.

§ 3. — Related Probabilities

The calculation of the probabilities must not be limited to the study of the phenomenon of pure chance.

Between pure chance and absolute knowledge extends an immense domain on which we can at least risk some timid steps.

If one considers a long series of tests, in order to fix the ideas of a game composed of a large number of hands, one is led to the classical laws of large numbers while assuming that these tests, or hands, are independent from one another.

The lone chance is in each of them or the chance and a cause which is in them completely independent. The classical laws of large numbers are the laws of pure chance.

One understands that it is interesting to study some more complex cases where chance and some other cause are not of an independent nature; where the chance and the other causes react on one another.

This will be the case, for example, for a game where, at each hand, the rules of play would depend on the previous total loss of the player. That which would be produced at each hand would not depend solely from chance and from fixed conditions known in advance, but also from the overall results of the previous hands.

The problems which do not assume the independence of the successive tests are relative to the "related probabilities," their resolution is generally very difficult.

If one considers a series of tests, one can imagine that these follow each other at equally timed intervals. When the number of tests is very large, each test happens in a unit of time (that is to say during an infinitely small time interval) and appears as an elementary phenomenon.

To say that the tests are independent reverts to say that those which can happen at a unit of time depends only on chance and on fixed variables or, more generally, on chance and on an explicit function of time.

When those which can happen in an element of time depend on more complex quantities, one says that there is *connectivity*.

I was able to make the theory of three classes of probability cases.

The first assumes that there exists a cause tending to diminish the écarts produced by chance and more effectively especially as these écarts are larger. That which could happen in each element of time depends on chance and on the result of the previous facts.

The second class assumes that that which could happen in each element of time depends on chance and on the previous maximum state, which is to say on the larger écart which happened previously. I have been led to the third class by the study of dynamic probabilities of which I will speak more later.

It is not necessary to confuse that which I call related probabilities with that which I named *mixed probabilities*. In the theory of mixed probabilities, of which I made known the first elements a few years ago, one considered some events which could tangle at each test, but the successive tests are independent.

§ 4. — Kinematic Probabilities

We say that a problem is relative to the kinematic probabilities when it consists of studying the displacements of a point or of a system, these displacements depend totally or in part on chance.

The simplest case is that where one considers the movement of a geometric point M, animated by a speed of which the magnitude is constant and for which the direction constantly varies randomly, every direction having equal likelihood.

We understand what is meant by these terms: at each thousandth of a second, for example, the direction of the speed changes brusquely and is at each instant independent of the previous direction and of the position occupied by the point.

One demonstrates that on average the distance from the point M to the initial position is proportional to the square root of time. This is a result analogous to those which we have found for the law of large numbers; the number of tests is replaced here by time.

The study of the movement of a point which moves at random is not only interesting from the speculative point of view, one such movement exists in nature, that is Brownian motion.

The difficulty of the same problem is much larger when the displacements due to randomness are combined with the given displacements, dependent, for example, on each instant from the actual point. One could meanwhile resolve the problem in the simpler case; it is especially interesting when there is conflict between the effect produced by randomness and the effect produced by the other forces of displacement.

The effect of randomness dominates in certain areas of space and the effect of the other factors of displacement in others; these areas are separated by a neutral surface where the forces are in equilibrium.

§ 5. — Dynamic Probabilities

We say that a problem is relative to the dynamic probabilities when it consists of studying the movement from one point or from an earthly system, the forces which act on that point or that system depends in total or in part on randomness.

One first studies the movement from a free earthly point, submitting to the action of a force for which the magnitude is constant and for which the direction varies constantly at random, every direction having equal likelihood.

We understand what is meant by these terms: at each thousandth of a second, for example, the direction of the force changes brusquely and it is, at each instant, independent of the previous directions and of the position occupied by the point.

It is a matter of calculating the probability for which the earthly point has a position and a given speed at the end of time *t*.

The difficulty of the question comes from that which the displacement of the point at each instant depends, not only from the direction of the force, but also from the speed attained, that is to say from the preceding movement. One finds thus from the presence of a problem for which the successive tests (here from the elements of time) are not independent, one finds oneself in the presence of a problem of "related probabilities" demanded by the special devices.

The velocity of the point grows, on average, proportionally to the square root of time and its distance from the point of departure grows, on average, proportionally to the 3/2 power of time.

One can likewise study the movement of the point in a resistant medium and some questions on the movement of the solid body.

The theory of kinetic and dynamic probabilities, that I call also *mechanics of randomness*, could give rise to some interesting research.

— Chapter XVII —

Randomness and Experiments

To verify by experiment the mathematical laws of randomness could seem ridiculous, the theory of the probabilities is based on rigorous reasoning, on an impeccable logic that is correct to the same degree as all mathematical analysis, and the material facts, whatever the number and the nature, can have no influence on the value of these deductions from the purely rational point of view.

But the calculation of probabilities, which is not more difficult in science, is maybe that which is understood by the least of men, and since it presents nothing which is visible nor tangible, as it conserves for many something of mystery, these conclusions are often only welcomed with a profound skepticism.

The experiment is the last means which remains to convince the non-believers. But the experiment is useful not only for that: without having the least doubt on the truth of a mathematical law, one could find interest in noting up to which point a material result could approach a result indicated by the pure theory.

The givens of a problem of calculation of the probabilities could anyway be hypothetical, the calculation transforms them by means that the experiment could be applied to them; the experiment could affirm or invalidate the hypothesis; it does not impact the affirmation or the invalidation of the calculation which is only a means of logical transformation.

The justification of the hypothesis could be useful in certain cases, for example when one wants to practically apply the mathematical theory of speculation.

More often than to justify some hypothesis, the experiment serves to verify some simple figures, and likewise in many of the cases it constitutes the only process which permits to obtain the value approached by a probability, not only when it is a matter of phenomenon of which the causes are unanalyzable by reason of our ignorance, but again when it is a matter of phenomenon which perfectly known causes lead to the practically impossible calculations.

Besides it is always interesting to compare the data given by the theory with those which are furnished by observation; from the similarity or from their dissimilarity one can always reach a useful conclusion.

Galton imagined an apparatus destined to give an immediate visible manifestation of the law of large numbers.

This apparatus that I cannot describe here, albeit being rather simple, could evidently prove nothing for or against one mathematical law; if these indications did not conform to the mathematical law, it must be concluded from this that its construction is defective and that randomness is not the only element which intervenes in the experiment.

The idea of giving an immediate visible and material representation to the law of large numbers is compelling, and the ingenious apparatus of Galton merits being better known.

R. Wolf had performed series of experiments materializing Buffon's problem of the needle.

The observations of Wolf, based solely on 5,000 experiments, gives for value of the ratio π of the circumference to the diameter the number 3.1596. The error does not reach one hundredth, which is small if one respects the relatively restrained number of experiments.

The roulette wheel permits the realization a little closer to perfection of the laws of randomness.

The roulette wheel contains 37 numbers; 18 of them correspond to the color red, to 18 others the color black; the 37th number is the zero. When the zero happens, the gambling house keeps half of the wagers and reimburses the other half. The probability of obtaining black is thus 18/37.

A first very easy verification is obtained by the series of at least five blacks. According to the theory, the average value of the number of tests that would be necessary in order to have black present itself at least five times in a row, is 69.52.

In 110,283 rounds, there occur 1,594 series of at least five blacks, being on average one series in 69.18 rounds.

The series of at least six black leads to an analogous verification; according to theory, there must be, on average, 144.95 rounds to produce one series of at least six blacks..

In 110,284 rounds, there are produced 762 series of at least six blacks; being on average, one series in 144.73 rounds.

The series of at least seven blacks must occur, on average, in 299.94 rounds, according to the calculation. In 100,109 tests, there have been 368 series of seven blacks or more; being one series in 299.62 rounds.

The average value of the number of rounds that it is necessary to play to obtain a series of at least eight blacks is 624, according to the theory. In 110,109 rounds, there occurs 176 series of eight blacks or more; being one series in 625 rounds.

Let's consider another kind of verification: the probability of red being 18/37, the average value, the probable and most probable number of occurrences of red in 518 rounds, is 18/37 × 518 or 252.

The difference between the number of occurrences of red in 518 rounds and the normal value 252 is the écart.

The mean value of the écart considered in absolute value must be, according to the theory, 9.07.

The experiment carried on for 51,800 rounds, divided into 100 sets of 518 rounds, gave 8.96, satisfying results if one considers the small number of sets.

Other verification: the probability of zero is 1/37. The probability for which, in 65 rounds, the zero does not occur a single time, is 0.168.

In 1,610 sets of 65 rounds, there have been 266 sets not containing zero; the probability for which zero does not occur will be, after the experiment, 266/1,610 = 0.165. The number given by the experiment is, as one sees, very close to the theoretical number.

The formulas contained in my *Traité du calcul des probabilités* [*Treatise on the Calculation of Probabilities*] would certainly permit other verifications; they permit also some resolution of the very difficult problems of roulette.

§ 1. — The Decimals of the Number π

Relative to the number π, one has asked for a long time a curious question interesting to the calculation of probabilities; could the decimals of that number be considered as succeeding at random?

One conceives what is necessary to understand by that expression: if one could calculate π with a million decimals, it would have the digit zero around 100,000 times, according to Bernoulli's law, the digit one around 100,000 times, the digits 2, 3, 4, 5, 6, 7, 8, 9 each around 100,000 times, if the decimals succeeded at random. It is very evident that they follow, in reality, after an inflexible law, but it is not certain that this law yields, on the whole, the results analogous to those that would be given by randomness.

It seems difficult to prove by these logical arguments that the decimals of π succeeded at random, we have thus resorted to the experiment in remarking that these conclusions do not carry a veritable truth.

One mathematician who showed a wonderful patience, M. Shanks, has calculated the number π with 707 decimals.

Are these decimals correct? They probably are without one being able to affirm it. The good faith of the calculator is beyond doubt, but the most skilled and the most conscientious could commit an error; it is meanwhile likely that the digits are correct.

The decimals

0, 1, 2, 3, 4, 5, 6, 7, 8, 9

present themselves respectively in amounts

74, 78, 74, 72, 71, 64, 70, 53, 72, 79.

As one can see, the digit 7 has disquietingly fallen behind and one could ask if the number π doesn't have a marked antipathy for this digit.

The normal value of the number of occurrences of each digit in 707 tests (while supposing that it is randomness which picks them) is 70.7, the difference between the number of occurrences of a digit in 707 tests and 70.7 is the écart. The écarts are here:

3.3 7.3 3.31.30.3

-6.7-0.7-17.71.38.3

The écart -17.7 corresponding to the digit 7 appears very considerable. The theory of large numbers permits calculating the probability for which, in 707 trials, an event with the probability 1/10 occurs with an écart equal or superior to ±17.7. That probability is 0.026, it is not exaggeratedly small if one considers that it is a matter of a worst écart.

To do the process of the worst écart doesn't convey the idea of the value of the whole.

To form such an idea is much more difficult, it is necessary to consider the analogue of the ellipsoids of probability in a nine-dimensional space.

The analysis of the subject leads me to the following conclusions: the sum of the écarts is characterized by one number.

In the case being studied, if the écarts are due to randomness, the characteristic number has for an average value 17.33; for probable value (the value which has any chance of being or not being exceeded), 17.15; for the most probable value, 16.82; for the mean quadratic value (square root of the average value of the squares), 17.88. There are not three chances in 100 that this number will be less than 10, nor three chances in 1,000 that it will be greater than 30.

In applying the formula to the decimals of π, the characteristic number has 16 for a value, value which has a probability of 0.16 of being exceeded.

Then if the écarts were produced by randomness, two times in five, they would have been much smaller in total than in the case of the number π, three times in five they would be greater.

That result is very satisfactory; *as far as the experiment permits to affirm it, the decimals of the number π follow each other randomly.*

Relative to the determination of the number π by chance, I have already cited the experiments of Wolf; I am now going to say some words of another procedure of which the idea comes wholly naturally to the mind when one studies the laws of large numbers. It is a question of the determination of π by the game heads or tails.

One flips a coin a thousand times at random, if heads presents itself 508 times, for example, we say that the écart is 8. More generally, we call écart the difference between the number of occurrences of heads and the normal number 500.

Let's assume that we do several series of a thousand flips; for each series we note the écart considered as an absolute value, that is to say without consideration of its sign and we calculate the square of the écart.

If we take the product of 2,000 by the sum of the squares of the écarts obtained in the different series and if we divide by the square of the sum of the écarts taking their absolute value, the result obtained must tend toward π when the number of the series increases more and more.

That result is very interesting, but it is devoid of all practical value; one hundred million series would likely give the common approximation $\pi = 3.1416$ and all new decimals would demand 100 times more series than the preceding, because the approximation depends on the square root of the number of series. Besides, if we wanted to obtain π with a great exactitude, it would be necessary to employ a series of more than a thousand tosses.

The theoretical result is very curious, that is why it is mentioned. Practically, its value is nothing like that of the other analogous processes that one could imagine and which require without doubt some number also very large.

— Chapter XVIII —

Speculation

In the theory of speculation we assume that the variations of the prices of government bonds, for example, are due to randomness and one studies the laws of these variations, that is to say that one looks for the probability for which, for a predetermined period, the price differs by a given quantity from the true actual price.

While saying that the variations of the prices are due to randomness, we are going to express that, because of the excessive complexity of the causes which produce these variations, all things considered, in reality, as if randomness alone existed.

In the theory of speculation, we take great care not to undertake the analysis of the causes which could act on the prices; that research would be in vain and would only lead to errors. It is precisely because we are going to totally ignore that which is possible to know, it is precisely because this study seems an inextricable complication, but which is, in reality, a very grand simplicity.

From the point of view of applications, the theory of speculation is very useful, since the results which provide the review of the data are in perfect accord with those which provide the calculation.

This concordance between the theory and the observation is equally interesting from the philosophical point of view; it proves, in effect, that the market for government bonds obey the law of randomness.

That result was expected, such a market, constantly submitted to an infinite number of influences, both variable and of a diverse nature, must finally behave as if any cause was not in play and as if the randomness was it alone. The results of the theory would be put at fault only if one cause was constantly in the same sense; in general, the diversity of the causes permit their elimination, the incoherence even of the market is its method, and it is because it obeys only any law it inevitably follows the law of randomness.

It is especially from the point of view of pure science that the theory of speculation has been used; it has been introduced by a necessary means, in the calculation of probabilities, the notion of time and of absolute continuity; it has made to be born the theory of the influence of probabilities and the theory of continuous probabilities; it has permitted to conceive the laws of randomness in a deeper and more general manner; it has expressed the law of large numbers in a pure and perfect form, free of the original errors, of the contingencies with discontinuous probabilities, from the approximations, from the asymptotes. If speculation did not exist, it would be necessary to imagine it in order to better conceive of the laws of randomness.

§ 1. — Variation and Instability

Let's put ourselves at a determined instant and, to fix these ideas, let's assume that it is a matter of the French government bond.

The market price at this instant is the price for which there are as many buyers as sellers; the buyers believe in the increase, the sellers in the fall. *The market*, that is to say the aggregation of all the speculators, believing neither in the increase or the decline, since, for the market price, there are as many buyers as sellers; it considers thus the market price as representing the true value of the considered bond.

But if the market believes neither in the rise or the fall, it must assume some movement of a certain amplitude is more or less probable.

The amplitude of the movements (as the market admits) is measured at each instant, for a determined value, by a single quantity, by a single parameter which we call *coefficient of instability*.

If that coefficient is large, the market admits that some strong movements of increase or decrease are probable, if it is small, the market admits that the variations will probably be rather weak.

We had placed ourselves at a certain instant in time; one minute later, for reasons which we should not undertake an analysis, we note another price and the market admits another coefficient of instability.

One minute later we will note another price and the market will acknowledge another coefficient, and so on.

At each instant there is thus to consider the true price and the coefficient of instability which measures the greater or lesser amplitudes of the future variations.

(It is evidently from the prices of the government bonds at maturity and it is necessary to correct the market price from the effects of the coupons and contangoes. In order to study these variations, I can only refer to my work on the theory of speculation.)

§ 2. — PRINCIPLE OF MATHEMATICAL EXPECTATION

The operations of the Bourse are subject to the *law of supply and demand* and as every speculator is free to initiate a trade, or its inverse, one can only admit a favorable or unfavorable trade of speculation *a priori* to one of the investors. A trade that would systematically favor one of the investors would not find a counter-party.

A trade can be, *a priori* neither advantageous nor disadvantageous; it is that which one expresses in saying:

The mathematical expectation of every trade is zero.

It is really necessary to realize the generality of this principle, it applies not only to the trades at fixed maturities and to option trades before expiring after a determined period, but also to all trades, including the complex, which would be based on the eventual movement of prices.

§ 3. — LAW OF PROBABILITY

Let's place ourselves, like before, at a particular instant, one quotes a certain price.

We take that price as a basis, that is to say that we will report all the future variations of this actual quoted price.

Thus we take the actual price as zero in order to have to consider only the écarts; for example, if the actual price on the French bond is 95 francs, the price 95.50 equates to an écart of 0 fr. c.50, the price 94.75 equates to an écart of -0 fr. 25.

The écart is positive when it corresponds to a rising above the actual price, it is negative when it corresponds to a lowering.

That definition of the écart being well understood, we can state the results to which the theory leads and to which the market unconsciously obeys.

The probability of any positive écart for a certain period is equal to the probability of a negative écart of the same amplitude for the same period.

For example, there is the same probability for which, at the end of 15 days, there is 20 cents of rise or 20 cents of decline.

The écarts follow the law of large numbers:

The ratios of the écarts which have the given probabilities of being exceeded are always the same, regardless of the future period being considered.

If the market, for example, acknowledges that there is one chance in two that the écart ±23 c. will be exceeded at the end of one month, it acknowledges that there is one chance in three that the écart ±23 × (30/21) = ±33 is exceeded in the same period and one chance in four that the écart ±23 × (37/21) = ±40 will be exceeded in the same period (refer to the example on page 97 [of the original text, page 67 this text]).

There is a 0.177 probability for which the double écart of the probable écart is exceeded. The probable écart was here ±23 c., the market acknowledges that there are about nine chances in 100 that at the end of a month, the écart +46 c. *is exceeded while rising and around nine chances in one hundred that the écart* -46 c. will be exceeded in declines.

When one knows the probable écart relative to a future period, one knows the probability of all the relative écarts of that same period.

All that which has been said relative to the law of large numbers, could be applied here; in particular:

The écarts are proportional to the square root of the time period.

(It is a matter of isoprobable écarts)

If the market actually acknowledges that there are two chance in seven for the écart ±32 c. to be exceeded at the end of one month, it acknowledges that there are two chances in seven that the écart ±32×2= ±64 will be exceeded at the end of four months.

(This last law is not a necessary consequence of the principles of the theory, it requires a principle generally called *the principle of uniformity*, which we assume and which besides the market nearly always verifies, the statistics prove it.)

All the écarts thus follow a similar law. Knowing the probability for which one of them is exceeded, one can calculate the probability for which another écart is exceeded.

The probabilities depend at the same instant from a single quantity: the coefficient of instability.

We have placed ourselves in a particular instant in time; five minutes later the price will be different, the market always follows the same laws, but with another coefficient of instability.

Some instants later, the price will have again changed, the market will always follow the same laws, but with a new coefficient of instability.

The existence of a unique coefficient at each instant, variable from one instant to the other, is one consequence of the principle of uniformity, a principle for which I don't believe useful to give the statement, because it is very delicate and could give a false interpretation instead.

This principle is nearly always verified by the market and we could, practically, consider it as correct. It is meanwhile interesting, from the rational point of view, to remark that most of the results of the theory of speculation are independent from that principle and that they are thereby a very high degree of generality.

Each time that a result could be based on the principle of uniformity, we will take care to mention it, it is a luxury which could oblige the scientific philosophy; for a practical study, this luxury will be superfluous.

§ 4. — The Game and Speculation

We understand the difference between the game and speculation: the player of games is certain to be able to play constantly under identical conditions, the speculator ignores the future variations of the coefficient of instability.

The speculator finds himself nearly in the same conditions as if he plays constantly at heads or tails, but with variable bets, following the whims of an imaginary that we call the market.

Another difference exists between the game and speculation: the player at a fair game cannot reasonably have the pretention of winning; the speculator, on the other hand, believes that he will very probably win.

The speculator ordinarily professes a profound disdain for the gambler who he considers as a sort of robot whereas he himself is thinking as having a power of superior prediction.

In reality, gamblers and speculators lose as much money as one or the other, there is only some difference in the manner.

The brokers are, for the speculator, the equivalent of the zero in roulette; if they didn't exist, the operations of the Bourse would be rigorously equitable by virtue of the laws of supply and demand.

It would also be as impossible to invent a combination making it inevitable to lose on a bond as it would be impossible to invent one making a certain gain.

§ 5. — First Problem of the Theory of Speculation

The most simple problem of the theory of speculation consists of finding the probability for which a given price holds for a fixed period, or, for which it returns likewise, to look for the probability for which a given price will be exceeded within a given period.

For example, one asks for the probability for which, at the end of 25 days, the price of the bond will be at least 15 cents higher than the true price.

The problem is analogous to that which we had resolved when it was a matter of some écarts in the game or the law of large numbers. We know that one has constructed some tables giving the values all calculated from probabilities.

For the particular case of speculation, the little table which figures, in the middle of my book on the probabilities is, practically, very sufficient and permits, to every question, an immediate response needing no calculation.

§ 6. — Second Problem of the Theory of Speculation

Another problem consists of finding the probability for which a given price will be exceeded before a given period.

For example, we ask the probability for which, before 25 days, the price of a bond is, at whatever moment, 15 cents above the true price.

One understands the difference between this second problem and the preceding: one is asking the probability for which the écart of 15 c. will be exceeded "at the end" of 25 days, one is asking now the probability for which the écart of 15 c. will be exceeded "before" 25 days.

This second probability is evidently larger than the preceding, because the price cannot be superior to 15 c. at the end of 25 days without having been earlier, whereas, in the interval of 25 days, it is necessary, at a given moment, to exceed the value 15 c. and discount then in a way to be less than 15 c. at the end of 25 days.

This second problem leads us easily back to the first:

The probability for which one price will be attained before a certain period is twice the probability for which this price will be exceeded in that same period.

If the market acknowledges that there are two chances in seven for which the price 15 c. would be exceeded at the end of 25 days (that is to say for which at the end of 25 days, the price would be superior to 15 c.), it acknowledges that there are four chances in seven that the price 15 c. would be exceeded in the interval of these 25 days.

If the market acknowledges that there is one chance in three that the price -10 c. would be exceeded at the end of 16 days (that is to say for which, at the end of 16 days, the price would be inferior to -10 c.), it acknowledges that there are two chances in three that the price would be exceeded in the interval of these 16 days.

The same tables of probabilities, which serve for the first problem, could serve for the second; just double the digits.

One could figure out, on speculation, some very interesting and very difficult problems for which the study could not find a place here.

For the numeric calculations, it is necessary to understand the probable écart relative to the period being considered; that given is deduced from the écarts of the options by the proceeds which will be described further on.

That which is necessary to fully understand, is that we calculate the probability that the market *actually* assumes, from after these *actual* givens, that is to say from after the actual écarts of the options. Nothing in our calculations is based on the previous results.

§ 7. — Probability Curve

The probability for which, at the end of a certain time t, the écart would be x, is a function of x that one could represent by a curve.

On the figure on page 107 [= p. 67, this text] are represented two curves of probability; the curve A B C was relative to the écarts for a certain period t, the curve A'B'C' is relative to the écarts for period $4t$.

It is necessary to state that the coefficient of instability could vary from one instant to another, the curves, for the same period, could, from one instant to another, become more splayed or more concentrated; but, in their variations, as I had remarked when it was a question of the law of large numbers, they reached the same shapes.

All these curves belong to that which the mathematicians call a similar family.

I again emphasize this important point: that which we understand by probability of an écart or more generally of any eventuality dependent on speculation, is the probability that the market actually attributes to it, that is to say the aggregate of the speculators.

The probability of any eventuality (for example, the probability for which a given price would be exceeded before a certain period), varies without ceasing, not only following the variation of price, but again following the variation of the coefficient of instability.

From the scientific point of view, one could generalize the theory of speculation and not acknowledge that the state of the bond would be characterized, outside of the price, by a unique coefficient; one could also reject the law of supply and demand. One obtains thus a theory which is superior to the purely scientific point of view, but which, from another point of view, has lost much of its interest, because it is no longer, to speak properly, a theory of speculation.

— Chapter XIX —

Transactions of Speculation

There are two principle types of transactions of speculation: futures transactions, option transactions.

These transactions can combine themselves *ad infinitum*, especially as one often trades several types of options.

The buyer of the futures limits neither his gain, nor his loss; he wins the difference between the price at purchase and the price at which he finishes his transaction by a sale. In other terms, he wins the value of the écart when that écart is positive, he loses the value of the écart when that écart is negative.

The inverse takes place for the future's seller.

Simple Option. — Options permit speculation while running a limited risk beforehand to a certain sum which is the amount or the value or the worth of the option.

The simple option is traded by the foreigner and, in France, in the speculation in merchandise.

Speculator A believes the forecast of an increase for a certain period *t*, and does not want to run the risk of unlimited loss, makes himself the purchaser of an option of an increase for the term *t*.

He pays a certain sum up front, called simple option, to speculator B who, himself, believes in a decrease.

Through the payment of that option, speculator A acquires the advantages of the futures investor, without running the risks.

On the expiration date *t*, he gains, like the futures buyer, if there is a rise above the actual price, but he will lose nothing if it is lower.

The buyer of the simple option loses at maximum the value of the option, his risk is limited to that amount; his gain, however, could be unlimited.

Options are treated the same during the falling such that the value is apparently equal to those of the rising option which has the same term.

To obtain the value of the simple option, it is enough to apply the principle of mathematical expectation: since anyone is free to buy or sell options, the transaction is not, *a priori*, either advantageous or disadvantageous:

The mathematical expectation of the whole transaction is null.

In writing that the expectation of the taker of the simple option is null, one obtains the value of that option if one knows the probable écart, or inversely.

When one knows the value of the simple option for a certain period, it is enough to multiply it by 1.688 to obtain the probable écart relative to that same period, that is to say the écart which has any chance of being or not being exceeded.

The probability of success of the holder of the simple option is independent of the period of the term, it has a value of 0.345.

At the same instant, one could trade some simple options for diverse terms.

The value of the simple option must be proportional to the square root of the amount of time which separates the terms.

The four-month simple option is twice that of the one-month simple option, when these two options are traded at the same instant.

The law of proportionality to the square root of time is based on the ordinary principles of the theory to which we add "the principle of uniformity" which has already been discussed. This principle has as a consequence the existence, at each instant, of a unique coefficient called the coefficient of instability characterized by only the amplitude of the future variations.

The coefficient of instability is the value of the simple option for the unit of time.

Straddle. — A speculator A, believing to predict a large movement in one direction or the other, and wanting to limit his risk, goes and buys a straddle or double option [or *straddle*], composed of a call option and a put option.

The probability of success in taking a straddle is 0.425.

§ 1. — OPTIONS, IN GENERAL

In the speculation on the securities, in France, one trades options similar to those which we are going to study, but of a nature a little more complex:

Speculator A, believing to predict the increase for a certain period *t* (for example, the end of the month), and not wanting to run the risk of an unlimited loss, goes and buys a call option for the period *t*.

He pays at first a certain sum called a premium to a speculator B who, himself, believes in the decrease.

Through the payment of that premium, the speculator A acquires the advantages of the futures investor, but without running the risks.

If the premium paid by A is the simple option, A acquires the advantages of the futures buyer at the actual price.

If the premium paid by A is less than the simple option, A acquires the advantages of the futures buyer, but at a price superior to the actual price, at a price *m*, called the price of the foot of the option [the exercise price, or "strike" price], established by the law of supply and demand and dependent on the premium paid, from the period of the term and of the instability which the market actually assumes.

If the premium paid by A is greater than the simple option, A acquires the advantages of the futures buyer, but at a lower price than the actual price, at a price *m* (negative), called the price of the foot of the option and dependent on the same causes as preceding.

The sum paid by A (in taking for unity a single security or 3 francs of 3% interest) is called the *montant*, the value or the importance of the option.

It is certainly evident that, for a similar period, the value of the strike is lower as the option is more substantial.

At the equality of importance of the option, the strike price is even higher as the term is longer.

Buyer A acquires by the payment of the premium the advantages of the futures buyer, at price *m*; his risk is limited by the *montant* of the option. If at the date of expiration, the price is higher than *m*, the buyer A wins the difference as if he had bought futures at price *m*;

on the contrary, if, at this time, the price is less than *m*, the buyer A loses nothing.

Through the payment of the premium, the buyer A is, in a way, assured against the eventuality of a loss.

It is in the use of employing the expression of *response* [expiration date] of an option instead of the period of the option and to say that one bought an option *of h*, in place of option of value *h*.

In reality, options do not pay up front, but only after the date of expiration. That detail has importance only from the point of view of compatibility; it is, for us, without interest.

Let's assume, for example, that the price of a French bond was 95 francs. Speculator A, predicting a rise for the end of the month and not wanting to run the risk of a futures purchase, buys the bond for an expiration of the end of the month at the price 95 fr. 15 for 0 fr. 10.

This amounts to saying that speculator A pays at first 0 fr. 10 per 3 francs of bonds (his risk is limited to that sum) and that he will acquire from that fact the advantages of the buyer of futures at the price of 95 fr. 05, that is to say at the price of 95 fr. 15 diminished from 0 fr. 10, value of the premium.

If at the moment of the call date, the price is less than 95 fr. 05, the speculator loses only 0 fr. 10, value of the premium.

From 95 fr. 05 it begins to gain on the purchase. At 95 fr. 10 it earns 0 fr. 05 on the purchase, and he has paid 0 fr.10 for the option, his total loss is 0 fr.05.

At 95 fr. 15, the speculator earns 0 fr. 10 on the purchase and loses 0 fr. 10 on the option, finding himself even.

Above 95 fr. 15, he earns proportionally on the increase, for example, at 95 fr. 40, he earns 0 fr.25 on 3 francs of bonds.

The price 95 fr. 15 is *the price of the option*; it is this price which figures on the quotes; the strike price is 95 fr. 05.

One sees that the transaction is similar to a futures contract executed at a price of 95 fr. 15, the price cannot drop below 95 fr. 05. When, in reality, the price is less than 95 fr. 05, the loss is 0 fr. 10.

The strike price is obtained while deducting from the price of the option the value of that option.

The écart of the option is the difference between its price and the price of the forward. That écart is obviously always positive; it is equal to 0 fr. 15 in the example considered.

The écart of an option depends on its value, of the duration which separates from the expiration and the assumed instability of the market.

It is evident that, for a similar term or date of expiration; the écart of an option is any larger than its value is lower.

It is again evident that the écart of an option of a given value is any larger than the term is longer.

When the instability of the market is added, the écarts of the option are necessarily augmented; we say that "they tighten."

The options that we are going to define trade only on the exchange, the buyers run a limited risk and the seller an unlimited risk.

If a speculator believes in the drop and wants to run only a limited risk, he sells forwards and simultaneously buys options; the result of these two transactions is a put option for which the value is equal to the value of the call.

In summary, the situation of the speculator who buys a call option for which h to the écart e for the term or expiration t is the following:

If, at the term of the expiration t, the price is less than strike, $m=e - h$, the speculator loses the amount h.

To deviate from this price, the loss of the speculator decreases and becomes null for the price e.

Above this price, the speculator earns proportionally to the increase; for example, to a given price c greater than e he earns the amount $c-e$.

§ 2. —LAW OF THE ÉCARTS OF OPTIONS

It is evident that, for the same maturity, the écart of an option must be even larger as its value is smaller; this is an immediate consequence of the definition.

If one confines oneself to that very vague ascertainment, one could be led to some bizarre results: I have already demonstrated for a very long time, that it would be possible, in properly determining the écarts of three options, to obtain infinite transactions giving a profit to all the prices.

The absurdity of this result permits to independently affirm from all notion of probability, that the écarts of an option for a certain maturity, one must be able to deduct the écart from all other options for the same maturity.

In order to obtain the law which links the écarts of the options to their value, use is made of the principle of mathematical expectation which is only the scientific expression of the law of supply and demand.

"The mathematical expectation of every speculation transaction is null."

Since anyone is free to buy or sell options, the purchase or sale of these options can be *a priori* neither advantageous nor disadvantageous.

If it was advantageous to sell options, if the écarts of those were systematically too great, no one would want to buy and the sellers would not find any counter-party; they would be obliged to reduce the écarts until that equilibrium of the offer and of the demand re-established itself.

The purchase of an option *a priori* is neither advantageous nor disadvantageous. It is that which one expresses in saying that

The mathematical expectation of the purchase of options is null.

This principle results in a transcendent formula which is the mathematical expression of the law of the écarts of the options and which permits the calculation of these écarts with the degree of exactitude that one would want.

One could replace that mathematical law by another which excessively delays little in these limits of the practical and which is expressible very simply in ordinary language.

One adds the value of the option and its écart.

One multiplies the value of the option by its écart.

One takes the product of the two results.

That product must be the same for all the options which have the same maturity.

Such is the convenient law of the écarts of options; one must note its extreme simplicity.

If the écart of the option of 0 fr. 10 on the 3% French bond is 0 fr. 30, what is the écart of the option of 0 fr. 25 for the same maturity?

It is simply, to resolve that question, to apply the law that we are going to state; the constant product is, for the option 0 fr. 10: (10 +

30) × 10 × 30 = 12,000. For the option 0 fr. 25 this product must have the same value; the écart of the option of 0 fr. 25 is thus 0 fr. 127.

I have made known the mathematical law of écarts of options in my *Traité sur la théorie de la speculation* [*Treatise on the Theory of Speculation*]. The convenient law has been exposed in the free rein that I professed at the Faculté des Sciences in 1910. For all the calculations relative to these questions, I can only refer to my work on the calculation of probabilities.

If one trades an option of a certain maturity, one can know the *probable écart* the market assumes for that maturity. From the knowledge of the probable écart, one deducts the probabilities of all the other écarts, as we learned.

In order to obtain the probable écart, one adds the value of the option and its écart.

One multiplies the value of the option by its écart.

One takes the product of the two quantities thus obtained, one divides it by two and one extracts the cube root.

The result is the *simple option* relative to the maturity considered.

We have seen that in multiplying the simple option by 1.688 one obtains the probable écart.

If, for example, the écart of the option of 0 fr. 10 on the French bond is 0 fr. 15 for the call date of the end of the month, the probable écart relative to that period is 0 fr. 21.

One could calculate the écart of an option for a given maturity, knowing the écart of another option for another maturity.

Meanwhile it is necessary to acknowledge then a principle that the preceding results do not require; the "principle of uniformity," to which however the market conforms to almost always.

It's enough, to resolve the problem, to note that the simple options are proportional to the square roots of time.

One example will make easy to know the method:

The écart of the option of 0 fr. 10 was 0 fr. 12 for 16 days, what must be the écart of the option of 0 fr. 25 for 36 days?

The simple option for 16 days is 0 fr. 11, from these givens; therefore the simple option for 36 days is

0.11 × (6/4) = 0 fr. 165, since the simple options are proportional to the square root of the time.

The simple option for 36 days was 0 fr. 165, the écart of the option of 0 fr. 25 for the same duration is 0 fr. 10.

In order to apply the laws of écarts of the options, it must be noted that the écart of an option is the difference between its price and the *true price* relative to the maturity or call date.

The true price is equal to the quoted price corrected by the effect of the coupons and of the returns. The difference between the two prices is generally negligible, but it is sometimes very sensitive, and if one doesn't pay attention, one could be led to some errors.

The process by which one determines the true price relative to a certain period is described in my work on the theory of speculation.

One could imagine all sorts of options, and it would be always possible to calculate the elements by the application of the principle of mathematical expectation.

Thus, against the prior payment of a certain option, the speculator could acquire the right to receive the difference between the highest and the lowest quoted price up to a certain date of maturity.

The value of that option would have to be four times that of the simple option having the same date of maturity.

Through the prior payment of a certain option, the speculator would have been able to acquire the right to collect the difference between the actual price, and being the highest quoted price, being the lowest quoted price up to a certain date of maturity.

The value of that option would have to be equal to the simple option of the same maturity multiplied by $\pi = 3.14159$.

§ 3. — FACULTIES

One trades on certain trading markets with some sort of intermediary between the futures operations and the options market: these are the "faculties."

Let's suppose that 60 francs are the price of a merchandise. Instead of buying one unit at the price of 60 francs for a given date of maturity, we could buy a faculty of the counterpart for the same maturity at 62 francs, for example. It is necessary to understand by that as any difference below the price of 62 francs we would lose only on one unit, thus that any difference below we earn on two units.

We would have been able to buy a faculty of the triple at 63 francs, for example, that is to say that, for any difference below the price of 63 francs we would lose on one unit, thus for any difference below that price we would earn on three units. One could imagine some faculties of the quadruple, and more generally of the faculties of the multiple order.

One also trades faculties in the falling markets, necessarily at the same spread as the faculties in the rising market at the same order of multiplicity.

In order to obtain the value of the spread of a faculty, one has recourse to the principle of mathematical expectation, scientific expression of the law of supply and demand.

Everyone is free to buy or sell faculties, the purchase or the sale of which cannot be *a priori* either advantageous or disadvantageous; *the mathematical expectation of the purchase of faculties is null.*

The application of this principle leads to the following results:

The écart of the faculty of the double is equal to the simple option of the same maturity multiplied by 0.68.

The écart of the faculty of the triple is equal to the simple option multiplied by 1.096.

The probability of success from the purchase of a faculty of the double is 0.394, the transaction will be successful four times in ten.

The probability of success of the purchase of a faculty of the triple is 0.33, the transaction will succeed one time in three.

When one knows the écart of a faculty, one knows by that the same simple option relative to the same maturity and, therefore, the probable écart assumed by the market.

If we admit the uniformity, the écarts of the faculties are proportional to the square root of the time that separates the maturities.

In purchasing a faculty of the double and in selling the futures simultaneously, one obtains, as the resulting transaction, an option at the high of which the importance is the half of the écart.

A speculator predicting a large movement in one sense or another could render himself simultaneously taking one faculty at the high and one faculty at the low; he wins if a large movement happens, he loses if the price varies little. The transaction is analogous to a straddle.

In France, on the exchanges, one doesn't trade faculties, one obtains a transaction analogous to the faculty of the double to the high while buying simultaneously futures and options. One obtains from the same a transaction analogous to the faculty double at the low while buying the option and in selling simultaneously futures in double quantity.

§ 4. — Complex Transactions

As one trades in futures and often up to three options for the same maturity, one could initiate at the same time some triple transactions and likewise quadruple transactions.

The triple transactions already come out from the number of those which one could consider as classic; their study is very interesting, but we cannot occupy ourselves with them here, let's confine ourselves to the notions on the double transactions.

One could divide them into two groups, following those which contain or not some asset.

The transactions containing some asset is composed from selling assets and from buying at premium or visa versa.

The transactions at "option contra option" consists in the sale of a large option and in the buying simultaneously of a small one, or visa versa.

The proportion of the purchases and the sales can vary to infinity; practically, the second transaction carries on the same number as the first or on a double number.

The sale of a future against the purchase of options of the same quantity has already been studied, it is the put option. The gain is unlimited at the fall, the loss is limited to the écart of the option.

The sale of a future against the purchase of options in a double quantity is analogous to straddle, it gives a gain in the case of a strong rise or of a strong fall, and produces a loss when the variations are not very large. The loss is limited, the gain unlimited.

The purchase of a future against the sale of an option in simple quantity or double is the inverse transaction from the preceding, the gain is limited, the loss unlimited.

The purchase of a large option against the sale of a small one in the same quantity produces, in the rising prices a gain equal to the difference of the écarts of the two options. In falling prices, the loss is equal to the difference of the values of the two options. The gain is limited, the loss is equally so.

Let's consider finally the purchase of a large option against the sale of a small one of twice the quantity, for example the purchase of one unit of a French 3% bond at an écart of 0 fr. 10 0 fr. at 25 against the sale of two units at an écart of 0 fr. 25 at 0 fr. 10.

Below some price -0 fr. 15 the speculator earns 0 fr. 20 on the options sold and loses 0 fr. 25 on the options purchased, his loss is thus 0 fr. 05.

From some price -0 fr. 15 the loss diminishes and at the price -0 fr. 10 the speculator earns 0 fr. 20 on the options sold and losing 0 fr. 20 on the option bought, finds himself even.

Below this price -0 fr. 10 he earns, proportionally to the rise, until the price of +0 fr.15 at which his gain is maximized; he gains then 0 fr. 20 on the options sold and 0 fr. 05 on the options bought; his total gain is 0 fr. 25.

The gain diminishes then proportionally to the rise; at the price of 0 fr. 40, the transaction gives a null result.

Below 0 fr. 40, the speculator loses proportionally to the rise; at the price of 0 fr. 70, he loses 0 fr. 30. His risk is unlimited, his gains limited.

Let us note, in order to finish, that all transactions of speculation have individually a null expectation, a transaction composed of several others, regardless of its nature and its complexity have equally a null expectation.

§ 5. — THEORY AND PRACTICE

We have assumed that the variations of the prices of the bond could be considered as the result of randomness; that hypothesis has been verified as long as there doesn't exist a cause having an importance absolutely preponderant and material, being of a continuous nature, being of an unbiased nature, in a sense clearly determined, as for example, the threat of a European conflagration.

Outside of that exceptional case, the market predicts the variations with an extraordinary ability of divination and it follows, at each instant, the law of randomness with a remarkable exactitude. The study of the statistics furnishes the proof of it.

These statistics present a difficulty which singularly complicates and which results from the variation of the coefficient of instability.

If, in order to simplify the calculations, one replaces the coefficient of instability variable by the mean value, one commits an error for which the theory permits besides some predicting the sense. If one wants to consider some of the variation of instability, the establishment of the statistics becomes painful.

The year 1911 has been particularly eventful; some very important external events have been printed in minutes, as to the price of the bonds, some convulsions and some brusque impulsions; also for that year, is it necessary to consider the variation of instability if one wants to prove that the market obeys at each instant the law of randomness.

For each day of the year 1911, I have calculated the mean écart for the following day, such that it unconsciously calculated the market, that is to say, following the price of the option for the next day.

If the market assumes, for example, that for tomorrow the mean écart must be 0 fr.08 and, if it is, in reality, 0 fr.16, I say that the mean écart must be 0 fr.10 and if it is in reality 0 fr.04 the relative écart is 0.4.

The relative écart is the écart related to the mean écart. By the consideration of the relative écart, the case of the variable instability leads to the case of constant instability. The relative écart must follow exactly the theoretical law of randomness.

Let's look up to that point it verifies that law:

The market assumes that the relative écart 0.5 has a 0.69 probability of being exceeded, the trial gives 0.67.

The market assumes that the relative écart "one" has 0.42 probability of being exceeded, the observation gives 0.39.

The market assumes that the relative écart 1.5 has a 0.23 probability of being exceeded, the trial gives 0.20.

The theoretical mean value of the relative écart is one, its observed value is 0.96.

One sees, by this example, that the market of the annuity is very exactly the law of the probability and that it predicts the amplitude of the variations with an extraordinary facility of foresight.

The theory of speculation presents thus a considerable interest from the point of view of reality, but it is from the point of view of rationality that we must adopt if we are going to appreciate all its beauty, it expresses, under the clearer and more knowledgeable form and the fundamental laws of randomness.

— CHAPTER XX —

PROBABILITY OF FUTURE EVENTS FROM PREVIOUSLY OBSERVED EVENTS

To forecast the future from after [ex post] the effects is the final objective of science. Can we expect to calculate probabilities by realizing such an ideal?

To predict the effects of randomness from after effects of randomness, such may be the only pretention of this calculation.

If a phenomenon has depended earlier on the random event and if, in the future, it must likewise depend on it, the calculation of the probabilities could be estimated, the uncertainty of happening throwing a new vagueness to the uncertainty of the future.

This uncertainty of happening is certainly going to change some things, which is the doubt which is sown and which will germinate if it has enough time; the randomness is based here on chance, the foundations are fragile, the future edifice will be less stable, especially as it becomes higher.

Based on the results of a hundred past tests to predict the results of a billion trials to come is of audacious temerity, it is that which the calculation expresses with a precision and an admirable simplicity.

It is not useful to insist on the importance of the subject treated in this chapter, it is evident. In reality, it is rare that one knows exactly the probability of an event; nearly always one deduces it from experience; nearly always, in consequence, one is led to study probabilities of future events from after observed events.

That which will survive assumes in a formal fashion that all the past or future tests are identical in the sense where one hears that term in the calculation of probabilities, that is to say identical to the nearly random effects.

Let's recall, in a few words, the law of Bernoulli, we will then deduce the inverse law of it.

An event has 0.3 for a probability; the mean value and at the same time more probable, the normal value, in some ways, of the number of happening of that event in 1,000 tests, 1,000 × 0.3 or 300.

If the event in 1,000 tests happens 307 times, the écart is 7, if it happens 295 times, the écart is -5, the écart is the difference between the observed number and the normal number 300.

More generally, if one eventually has for probability exactly r, the normal value of the number of happenings of that event in m tests is mr.

If the event, in m tests, produces in number $n=mr+x$, one says that the écart is x.

When the number of tests varies, the isoprobable écarts are those which have equal chance of being exceeded.

For example, the écart which has one chance in three of being exceeded in 100 tests and the écart which has one chance in three of being exceeded in 700 tests are isoprobable.

When the number of tests is very large the isoprobable écarts are proportional to the square root of the number m of tests, they become thus smaller and smaller relative to m when the number of tests grows.

This is the law of Bernoulli that one can call *direct*, we deduce the *inverse* law of it.

§ 1. — Inversion of the Law of Bernoulli

We ignore the probability of an event, but one knows that it happens 300 times in 1,000 identical tests; the probability of the event for a new test identical to the preceding is necessarily in the neighborhood of 300/1,000 = 0.3.

If one knows that the event happens three million times in ten million tests, the probability of the event is necessarily around (3,000,000/10,000,000) = 0.3 and one could be much more positive relative to the exactitude of that number 0.3 than one would be in the case of the preceding example. In effect, after the direct law of Bernoulli, if the event had a probability very different from 0.3 (for example, 0.29) or could hold as practically impossible that this event could happen three million times in ten million tests.

If the event happens three times in ten tests, we would adopt maybe again, for lack of a better, 3/10 = 0.3 for the probability but without prejudice.

One conceives the sense of the law of Bernoulli called *inverse.*

The ratio of the number of happenings of an event to the total number of tests approaches the probability of that event especially as the number of tests is larger.

That law is also evident. One could present it under another form: the number 0.3 is, in the preceding example, the apparent value of the probability or the observed probability. If the number of tests increases more and more, the *apparent probability* differs less and less from the real probability.

It is not sufficient to know that the difference between the real probability and the apparent probability diminishes when the number of tests grows, it is necessary to know from after which law she diminishes if one wants to make an idea of errors to fear when one adopts the apparent value, in the case where one is ignorant of the exact value.

When the number of tests is very large, that law is deduced from the law of large numbers; we will not study it in detail, we will content ourselves by a simple preview.

The error committed in substituting the apparent probability by the exact and unknown probability, varies in general, inversely to the square root of the number of tests.

While augmenting the number of tests, one therefore approaches more and more toward the exact value, but very slowly.

If for example, one knows the probability with two exact decimals, it would be necessary to centuplicate the number of tests in order to obtain it to three decimals.

In summary, when the unknown probability of an event must be deduced uniquely from experiments performed in some identical conditions, one adopts for the value of that probability the ratio of the number of happenings of the event to the total number of tests.

The error committed thus varies, generally, inversely to the square root of the number of tests.

I insist again on the point, that the tests are assumed to be identical and very numerous.

§ 2. — PROBABILITIES OF FUTURE EVENTS FROM THE OBSERVED EVENTS

One event has been produced 300 times in 1,000 tests; in the ignorance where we are from the exact value of the probability of that event, we adopt, after Bernoulli's inverse law, the ratio 300/1,000 = 0.3, this is the *apparent* (or observed) *probability*.

If one must try 10,000 new tests, the mean value, probable and more probable, the value, in some normal way, of the number of happenings of the event is $0.3 \times 100{,}000 = 3{,}000$.

More generally, if an event happens n times in m tests, one assumes for the value of the probability the ratio $n/m = p$, this is the apparent probability.

If m' new tests must be held, the normal value of the number of events which would produce in these m' tests is $m'p$.

But it is very necessary to note that here it is a matter of an *apparent normal value*. As the probability p is erroneous, the apparent normal value $m'p$ is equally erroneous.

The error committed on the apparent probability p varies, generally, inversely to the square root of the number of previous tests; for which the committed error on the apparent normal value $m'p$ varies directly by the number of future tests and inversely to the square root of the number of previous tests.

In this preceding example, if 10,000 new tests must be held, the apparent normal value of the number of occurrences of the event is 3,000. If the event should happen 3,025 times, one says that the *apparent écart* is 25.

More generally, if the event with apparent probability p must happen $m'p + z$ times in m' future tests, one says that the apparent écart is z.

The direct law of Bernoulli gives a strong simple idea of the manner which raising the écarts when the number m' of future tests are added: the isoprobable écarts are proportional to the square root of the number of tests. But that law assumes a formal manner that the probability for each test is exactly known.

If that condition is not filled, if the probability is given by the experiment, the law of Bernoulli is no longer exact.

How does one know? When the probabilities are known the écarts depend only on the caprice of randomness in the future. When the probabilities are given by experience, the écarts depend more on the caprice of randomness in the past; they depend doubly on randomness, they must be larger than in the first case, the truth of this fact is evident.

The caprices of randomness in the past tests have distorted the actual givens in an indelible manner, all our calculations relative to the future would keep the imprint of that original fault.

Let's try, without any calculations, to realize the manner for which they could raise the apparent isoprobable écarts when the number of future tests will increase itself indefinitely: I recall that one calls isoprobable écarts those which have equal chance of being exceeded; for example, the écart which has one chance in four of being exceeded in 10,000 new tests and the écart which has one chance in four of being exceeded in 14,000 tests are isoprobable. In the following, I will delete as regular the word isoprobable, it will be implied everywhere.

In adopting the apparent probability $n/m=p$ one commits an error, that error is unknown, we know that it varies in general, inversely by $\sqrt{m}$ and that it is certainly very small; regardless, it exists.

It exists and it is of a systematic nature, that is to say that it must reproduce identically to itself at each new test, chance not having, in the future, any influence on it.

Its effect will be proportional to the number of tests. In which sense does it reproduce? We don't know since the error can with equal chance be positive or negative; the effect of the error is thus to increase the absolute value of the écart proportionally to the number of new tests.

Thus, when the number m' of future tests increases, the apparent écarts increase under two influences: the first is that of randomness, it is this alone which would exist if the observed value p was exact; therefore by this first influence the écarts increase, as we know, proportionally to the square root of the number of the future tests. The second influence is that of the systematic error committed by adopting the observed probability p; by it, if it were alone, the apparent écart would rise proportionally to the number of tests. The systematic error was very small, the effect of the second influence is negligible if the number *m'* of future tests is very small relative to the number *m* of past tests, but if the number *m'* increases by more and more, the effect of the second influence was proportional to the number of tests thus that the effect of randomness is only proportional to the square root, the second influence will happen to equal the first, then it will surpass it and will finish by dominating as much as one wants.

The law of Bernoulli which would be relative only to the first influence isn't thus more applicable when it is a matter of observed probabilities, at least that the number of tests are not very large next to the number of future tests.

By which law must one replace the law of Bernoulli in the case of the apparent écarts?

From the law of Bernoulli the écarts are proportional to $\sqrt{m'}$. The apparent écarts are proportional to $(\sqrt{m'})*()*m+m')/m)$. That formula shows that the écarts are doubled when the number of future tests is triple the number of the past tests; they are tripled when the number of future tests is eight times larger than the number of past tests.

That last formula made known the importance of the doubt that is introduced by the uncertainty of the past; if that uncertainty does not exist, the last radical would have one for a value.

The probability of an apparent écart *z* (it is no longer a matter here of isoprobable écarts) is one function of *z* that one could represent by a curve. The curves of the apparent écarts are the same, from the geometrical point of view, as the curves of the true écarts (page 107) [page 67, this text], but, in the first case, the deformation of the curve is much more rapid; the curve flairs much more quickly when the number of future tests increases.

I have shown, in my *Traité du calcul des probabilités* [*Treatise on the Calculation of Probabilities*] that the theory of probabilities of future events, from after the observed events, is a particular case of the theory of related probabilities of which I had spoken previously.

§ 3. — Probabilities "a posteriori" [from the latter]

We have seen that if an event produces itself *n* times in *m* trials, one must adopt for the value of the probability of the event the ratio n/m if *n* and m-n are large numbers.

If n and m-n are not large numbers, what value must one adopt?

In the absence of all other pieces of information, it seems natural to adopt again the value of n/m, but without prejudice and without wanting to claim that it is necessary.

The adoption of this ratio n/m constitutes a hypothesis which presents certain advantages: the mean value of the error thus committed is null, the mean value of its square varies in inverse relation to the number of trials; thus this hypothesis, by its simplicity, permits a very elementary theory. These advantages do not permit the adoption without prejudice the ratio n/m; besides, we are going to study the subject from a perspective much more deeply, from Bayes and Laplace.

Without knowing of the precise manner of the probability of the event, one must have on that probability some more or less vague notion, expressed itself by some probabilities.

For example, without knowing the probability of an event, one might have some reason to assume that it has one chance in three of being less than one half and two chances in three of being greater than one half. That given *a priori* constitutes a relative understanding.

The data *a priori* could be more or less complex; mathematically, in order to obtain the greatest generality, one considers them as whatever and one expresses them by an arbitrary function. These data *a priori* or initials are often called hypothesis *a priori* or initial hypothesis.

These initial data or data *a priori* on the probability of the event being known, one is going to know that this event happens n times in m tests.

This new fact, certain by itself, teaches us absolutely nothing certain about the probability of the event; but, like the data that we possessed previously was equally uncertain, the two relative understandings, the initial knowledge and that which we bring the new fact, goes, in some way, to amalgamate to produce some relative knowledge *a posteriori.*

The relative understanding *a posteriori* on the probability of the event depend thus from two things: from the knowledge *a priori* and from the existence of the new fact.

In order to fully comprehend what will be these relative understandings *a posteriori*, we are going to assume that the number m of tests is at first very small and that it increases then until infinity.

If m number of tests is very small, the new fact doesn't have great importance and the relative knowledge *a posteriori* retains the imprint of the initial data.

But, as the number of tests increase, the initial hypothesis, contradicted by the multiplicity of the adverse tests, constantly denied by the facts, is going to lose its influence more and more, soon it will not keep any more vestige and, that which would have been the initial hypothesis, the results will be the same; it will be consistent with the inverse of the theorem of Bernoulli:

When an event produces itself n times in m tests (n, m and $m - n$ being very large numbers), its probability is very close to n/m.

The error committed while adopting these data n/m varies, generally, as the reciprocal of the square root of m number of tests.

When the number of tests is not very large, the understanding *a posteriori* of the probability depends on the knowledge that one has *a priori* of that probability. In general, the knowledge *a posteriori* does not permit (outside of the case of large numbers) the adoption of a unique number, preferable to the others in order to represent the probability.

The knowledge *a posteriori* gives a figure for the mean value of the probability of the event, another figure for the probable value, another for the most probable value. The same equality of these three values will not be sufficient, strictly speaking, in order to asses them as the ratio n/m, when the number of tests is large.

§ 4. — Hypothesis of Bayes

Very often one knows nothing *a priori* about the probability of an event. Then it is natural to assume that all the values of that probability included between zero and one have, *a priori*, equally likelihood.

That hypothesis leads to the following results: the event having occurred n times in m tests, the most likely value of the probability of that event is n/m and the mean value of that probability is $(n + 1) / (m + 2)$.

The hypothesis of Bayes leads thus very nearly to the same results as the hypothesis previously considered which assumed the adoption of the ratio n/m, in the uncertainty where one is in order to adopt another value.

The consequences of the two hypotheses are the same; the first is certainly preferable, from the rational point of view. The advantage of the second resides in its extreme simplicity.

It is indispensable to fully comprehend on what basis is founded the theory of the *a posteriori* probabilities if one isn't going to expose oneself to using it for incorrect applications.

It assumes, after all, and in a formal manner, the identity of all the tests.

It assumes further that the hypothesis *a priori* is unique and initial data.

This last point requires an explanation; it has great importance. For greater clarity, let's assume that the m tests are successive, and we are closely studying the method of the classical theory:

The knowledge of the hypothesis *a priori* is going to combine with the knowledge of the first test in order to produce a first understanding *a posteriori.*

This one could be considered as an understanding *a priori*; it will combine with the knowledge of the result of the second test in order to produce a second piece of information *a posteriori.*

This one maybe considered as a piece of information *a priori*, it will combine with the knowledge of the result of the third test in order to produce a third piece of knowledge *a posteriori*. Generally speaking, the knowledge *a posteriori*, after the mth test, might be considered as a knowledge *a priori* for the following tests.

It is this which acknowledges the theory, it is not that which one necessarily assumes in many of the cases.

In many of the cases, as a result of the nature of the question being studied, the results of the first test absolutely change the hypothesis *a priori* that one assumes before undertaking the second test.

The classic theory assumes only one arbitrary hypothesis, at the beginning, while it should be possible, in some manner, to interpose a new hypothesis between each test.

An example will make this slightly abstract thought very understandable.

A scientist has a vague idea that an experiment must succeed, meanwhile he is not completely persuaded of it: he does three experiments, all three succeed; he could very justifiably believe himself in possession of a certain scientific fact before necessarily replicating it in future trials..

After the first experiment, the doubt of the scientist is nearly transformed in conviction; that first success doesn't have only the importance of an accidental fact, it implies the possibility of a constant law.

The hypothesis *a priori* before the second test must not only be a consequence of the hypothesis *a priori* before the first test, it must also be a consequence of the new idea that that success must imply.

After the general theory, one would not consider that last consideration and the third success would not prove a big deal.

In summary, the classic theory assumes an initial idea and the facts could transform it, but it does not assume that these facts can give birth to new ideas.

Each time that a fact will be of the nature of introducing a new idea, the theory will not be applicable or at least it will modify it.

One could base on the study of probabilities *a posteriori* the theory of the probabilities of the future events from after the observed events. One obtains nothing which will be new for us when the number of completed tests is large, since, in that case, the result *a posteriori* is independent of the hypothesis *a priori.*

If the number of future tests is itself very large, the law of Bernoulli is no longer applicable to these tests, we have seen. The result *a posteriori*, in order to be independent of the initial hypothesis, doesn't contain less doubt; that doubt, as small as it is in importance for only one new test, takes on in the long term a profound influence.

The doubt, as minimal as it is, changes the nature of the problem; in the long run, it completely changes the result.

While expressing in a fashion a little incorrect, but colorful, one could say that the future is especially uncertain in that it is much longer and that the past is very uncertain itself.

§ 5. — CONDITIONS OF IDENTITY OF THE TESTS

For the preceding problems, we assume that the tests are identical. Or, in many of the cases, one ignores that and one demands precisely of the calculations of the probabilities of giving some indications which could permit to consider that identity as more or less likely.

One conceives that this type of question presents a great interest from the point of view of the applications.

One event happens 500,000 times in a million tests; what is the probability that it will happen at the next test?

If the identity of the tests is not recognized, we know absolutely nothing.

How to recognize the identity of the tests?

The idea which comes all naturally is to split the million tests into several groups and see if the results given by the different groups are compatible, to the close errors that could have occurred by chance.

If the tests are identical, one should assume 1/2 or 0.5 for the value of the probability of the event and there is not a chance in five million for which that probability would be superior to 0.503 or inferior to 0.497.

We will divide the million tests into a thousand groups of a thousand tests; if for one group, the event happens 490 times, the observed probability for that group is 490/1,000 or 0.49. The difference between the generally observed probability 0.50 and the probability 0.49 is an écart.

If one considers the écarts for all the groups, they must follow at least nearly the law of large numbers; otherwise the probability is not the same as all the tests, these are not identical.

If the tests follow in a determined order, it is very necessary to keep from changing their order, the consideration of the observed probabilities for the different groups could, in certain cases, lead to interesting results, finding periodicity, for example.

If the tests are themselves divided into groups, it is necessary to keep the mix under pretext to obtain the equal groups. If for example, one knows a statistic for each of the 86 unequal numbers, but characteristically; the calculation permits the consideration of the inequality of the groups.

When there are only two groups to consider, one could employ a due process according to Laplace: let's assume, for example, that one has observed that in m births in France there are n males, England on m' births there are n' male births; m, n, m', n' were of very large numbers, the ratios n/m and n'/m' must be very nearly equal if randomness alone produced the difference. For the formula of Laplace one could calculate the probability for which that difference has a given value if it is due only to randomness and in conclusion, in certain cases, that very probably a cause, other than randomness, has entered the game.

The calculation of the probabilities does not permit to affirm, in the rigorous sense, the existence of a cause having acted outside of randomness. But these arguments, so as not to don the imperative form, are not less peremptory in certain cases: "There is not one chance in three million that randomness has produced such an event," he says. He is a statistician in the end.

§ 6. — Male and Female Births

The study of statistics is quite out of our program; the study of mortality has been brilliantly treated in one of the preceding volumes of this collection; we limit ourselves thus to acquire some notions on a very interesting subject which, for two centuries, has given place to very numerous observations.

It is a matter of the study of the ratio of the number of male births to the number of female births.

For the purpose of simplification, we will declare from now the entire result that is necessary to remember:

"The ratio of the number of male births to the number of female births is constant; it has for a value 1.05."

We understand that the constant is only relative and that the ratio must vary somewhat.

It was Arbuthnot who, in 1710, did the first research on the subject; he was amazed by the consistency of the ratio and therein saw one of the more evident proofs of the permanence of the laws of nature.

These statistics were relative to births in the city of London, between the years 1629 and 1710; the mean value of the ratio was around 18/17 or 1.06. The extreme values were 1.14 and 1.01.

These results were submitted to Nicolas Bernoulli who, far from being marveled by them, claimed that they were nothing except very normal.

While admitting, he said, that the probability of the birth of a boy was 18/17, the differences were explained by assuming that randomness alone produced them; these differences could be considered as accidental.

But precisely, as Moivre points out, causing the admiration of Arbuthnot, it is that one can assume a constant probability and that randomness explains the rest. Far from diminishing the value of the discovery of Arbuthnot, the criticism of Bernoulli adds considerably to the interest in it.

For two centuries, one has studied the variation of the ratio in each country, following the age of the parents, following their profession, following the climate, following heredity, following the nationality, etc.; the ratio varies little. Very rarely, it descends to 1.02 and, very rarely, it is greater than 1.08.

If one considers the entire country of France, and if one calculates the quinquennial averages, one sees that the ratio considered is always between 1.068 and 1.04 inclusive. Its actual value is 1.044. The variations of the ratio could be considered as a bit accidental, randomness nearly sufficing for the explanation.

For several years, one frequently used the neologism "masculinity" to designate the number of male births per 100 female births. The masculinity is always in the vicinity of 105 and nearly never exceeds the limits of 102 and 108.

§ 7. — Births in the Same Family

One very interesting question, of which the idea comes totally naturally to mind, consists of finding if, in the same family, the previous births don't have influence on the births to come, from the point of view of the genders of the infants. One family having had four daughters, that fact, does it augment the probability that the fifth infant will be a daughter? Does it diminish that probability? Does it leave it indifferent?

Let's assume, in order to simplify, that, *for the entire* population, there are born as many boys as girls. For the entire population, there will be one chance in two for which a birth will be masculine, one chance in two that she will be feminine. "In the aggregate," the genders would produce themselves as if one played their arrival as heads or tails, or as red and black.

To say that in aggregate there are born as many boys as girls does not imply necessarily that in each family it will be thus. While exaggerating to the extreme, one could assume that half of the families had only boys the other half only girls; the general proportion would be maintained.

Let's consider the families having five children; if the genders are produced at random, as if one played the event like heads or tails, the probability of five feminine births would be 1/32; the probability of five male births would be equally 1/32. The probability of four births of the same gender and of one birth of the opposite gender would be 10/32. The probability of three births of one gender and two births of the opposite gender would be 20/32.

If thus the genders were produced at random, without any one birth being influenced by those which preceded it, approximately one family in 32 would have five girls, one family in 32 would have five boys, ten families in 32 would have four boys and one girl or four girls and one boy; 20 families in 32 would have two girls and three boys or three girls and two boys.

If these proportions are not very nearly verified in reality, one must conclude that randomness is not alone in determining the gender of the children in the family and that anther cause is certainly in play.

If, for example, five births of the same gender happen one time in 40, instead of one time in 16, it would be very necessary to admit that nature was following a sort of law of compensation leaving more probable the birth of a boy when there were already several girls, and vice versa.

If, on the other hand, five births of the same gender happen, for example, one time in 12, it must be admitted that each couple, following the organism of spouses, is predestined rather to have girls or instead boys.

Some statistics have been recently established by M. Lucien March, one could deduce from them this general result: "The genders of babies from the same family one after another is random."

Thus erasing the common prejudice from after which the gender of a child to come would depend, for a large part, upon the gender of their older siblings.

If one examines the statistics more closely, one sees that the previous births influence those which follow them, but very little.

The probability for which the two first infants of a family would be of the same gender would have to be 0.500 if randomness were the only cause: the statistic given is 0.506.

The birth of a first boy renders a little more probable the birth of a second boy, the birth of a girl leaves a little more probable the birth of a girl, there is a tendency toward the production of the same gender; but that tendency, which seems to accentuate lightly toward the following birth is always very fallible.

One could thus say that, for the infants of the same family, the genders follow one another at random.

It is quite another thing when it is a matter of simultaneous births; if randomness produced the genders of the twins, one time in four, on average, they would be both of the male gender; likewise one time in four, on average, two coins tossed at random fall on the side of heads. In reality, one time in three the twins are both of the male gender, one time in three they are of the female gender, one time in three they are of opposite genders. It is necessary to conclude from this that one singular cause, acting in a very delicate manner, tends to give the same gender to twins.

— Chapter XXI —

Errors of Observation

One cannot exactly measure a grandeur. Perfection is not in the earthly possibilities, each measure has an error and it will always be thus, despite all the achievable progress.

But if one cannot eliminate the errors, it is nonetheless admirable that we achieved, a century ago, to demonstrate that it follows completely the same law when randomness is the only cause.

The theory of errors is useful in all the sciences of observation, each time that, by some measure, one must obtain a value; it is, for example, of a continual usage in astronomy and in ballistics.

Where one measures several times, with care, a magnitude (in order to set this idea, a length), one obtains some different values.

It is necessary to conclude that each of these values do not represent exactly the magnitude and that each of those contain an error.

One calls *error* of an observation where one measures the difference between the exact value from the measure quantity and the value obtained at the considered observation.

The exact value is nearly always unknown, so that the errors are too.

One always assumes, in the questions that we will consider, that the errors are independent of the magnitude of the measured quantity.

Systematic and accidental errors. — It is necessary to distinguish the systematic or constant errors from fortuitous or accidental errors.

The systematic errors hold to a constant fault of the appearance which serves to measure or from the manner by which one measures.

For example, if one measures a length with a too long measuring stick, it results in a systematic error.

Accidental errors are, conversely, due to a bunch of variable causes from one observation to another, but compensating one another on average, in such a way that one can consider as if due to randomness.

In the following, we will suppose that there are no systematic errors or that, if they do exist, one knows their values, in a way that one could correct the obtained values.

Realistically, this hypothesis is unfeasible, it exists always from systematic errors, but it could be very fallible.

§ 1. — Laws of Errors

One calls the law of errors a relation which makes known the probability of a given error. It is by a geometric representation that the notion of a law of error becomes very clear.

Taking the exact value of the measured quantity as the origin, taking the errors as the horizontal axis and the corresponding probabilities as the vertical axis, one obtains the curve which represents the law of errors.

We could assume that the positive errors have the same probability as the negative errors of the same amplitude; thus the error curve is symmetrical.

Let us realize the general form that could reasonably affect a curve of error: the little errors are more probable than the large, as the null error will correspond to the highest point of the curve and that which would approach from the x axis as the error x will be greater.

The error curves would thus present the form of a bell more or less splayed like those which are represented on page 107 [= page 67 this text].

The area contained between the error curve and the x axis always has a value of one, since it represents the sum of probabilities of all the possible errors.

§ 2. — Exponential Law

There is a law of error of special importance, it is based on a simple and plausible hypothesis, and the evidence nearly always verifies it; the hypothesis which leads there, from Laplace, is the *hypothesis of infinitesimal errors.*

Let's try to create an idea of the manner by which an error is produced:

An error is the result of an infinite amount of infinitesimal errors due to an infinity of little causes that we are not even able to try to analyze.

These little causes are independent from one another and, since it is randomness which directs them, it is a matter sometimes in one sense and sometimes in another.

Such is the hypothesis of infinitesimal errors; it is, as one sees, simple and plausible, and one conceives that nature nearly always verifies it.

The hypothesis is sufficient to impose a unique analytic form to the law of error with, as a lone possible variant, a unique coefficient that characterizes more or less the precision of the observation.

We are going to see that the theory of the errors easily leads to the theory of grand numbers:

Let's imagine a player A; we match each of the hands of his game to each of the infinitesimal causes; one cause will produce a little effect or infinitesimal error in one sense or another that we assimilate as a gain or loss for the player; as a gain, for example, when the infinitesimal error is positive, as a loss when it is negative.

The causes are undifferentiated in one sense or another, the game is equitable.

The causes are very numerous, the hands that comprise the game are very numerous. The causes are independent, the hands of the game are independent.

The error is the result of the infinitesimal errors. Similarly, the total gain or loss of the player are the result of the gains and losses realized by all the hands.

The research of the law of error leads us thus to the following problem:

A player must play a very large number of hands at an equitable game; what is the probability that he will win (or lose) a given amount?

That is the problem of the theory of large numbers for which the solution is expressed geometrically by the curves represented on page 107 [page 67 this text].

There is no worry about the simultaneity of all the hands; provided that they are independent, it matters little that they be successive or simultaneous. The curves that express the gains or losses of the player always belongs to the same family, whether the hands are identical or not, successive or simultaneous.

The law of error will be thus represented geometrically by these same curves, their equations will differ only by a coefficient.

That coefficient is called *precision*. For example, the two curves from page 107 [page 67, this text] are the exponential curves of error; the precision, in the case of the curves A B C, is twice the precision in the case of the curve A'B'C'; the probability for which the error will be greater than $\pm a$ in the first case is equal to the probability for which the error will be greater than $\pm 2a$ in the second.

The exponential law of error is, in reality only the law of large numbers, not being, for us, a new study; it is sufficient for us to remember the subject treated in chapter XII and to replace in every question the word écart by the word *error*. Thus, the fundamental theorem, expressed on page 99 [page 62, this text], will be expressed thus:

The ratio of the errors which have some given probability of being exceeded are always the same.

If one knows the probability for which a certain error would be exceeded, one could deduce from it the probability for which another given error would be exceeded.

The probable error is, by definition, that which has one chance in two of being exceeded (error of excess or of shortcoming).

If, for example, the probable error on the measuring of a length is 1 mm, that signifies that there is one chance in four that the error will be greater than 1 mm short, one chance in four for which it will be less than 1 mm short, one chance in four that it will be less than 1 mm in excess and one chance in four that it will be greater than a millimeter in excess.

The probability for which the error (more or less) exceeds twice the probable error is 0.177. The probability for which the error exceeds triple the probable error is 0.043.

If the observations are made with much care, if the measuring sticks are very precise, the probable error is very small and all the others are proportionally.

If the observations are made with less care, the probable error is greater and all the others are proportionally.

The ratios are constant; the law is always the same.

The greater or lesser exactitude of the measures is expressed by a single value which is, for example, the coefficient of precision, or again the probable error. When it is a question of a game, we have likewise a characteristic value, and, as soon as it is a matter of speculation, we have to consider a coefficient of instability.

All these questions are analogous; they are, in reality, only some different forms of the law of large numbers.

The expression of the exponential law comes from the analytic form of the law. The expression of the normal law is self-explanatory; the hypothesis of infinitesimal errors is so plausible that one could consider as abnormal the case where it is not realized; it is however that which proves the experiment.

The same law of error is often called the law of Gauss.

— Chapter XXII —

Principle of the Mean

One has measured n times the same size (to set the idea, a length); one has obtained the values u_1, u_2, u_3... u_n. It is a matter, in the uncertainty as to where one is from the exact value of the measured quantity, to deduct from the obtained value the better value to adopt for the measured quantity. One assumes that the law of error is given and that it is the same for the n observations.

The problem thus proposed is generally unsolvable, due to its grand pretention. In general, there does not exist a value which, logically, imposes itself as being preferable to all others.

Under what conditions would a certain value, u, be adopted as logically preferable to the others?

It would be necessary at first that it give the greatest probability to the observed fact, that is to say to the obtaining of the values u_1, u_2, u_3... u_n; but that is not sufficient: a maximum of probability must not be indisputably preferred as if it is the center of symmetry, it must thus be a second condition: two values equidistant from the value u correspond to the maximum must give some equal probabilities to the observed facts.

This last condition is of extreme rigor. One conceives that outside of exceptional cases, the value u obtained by the first condition does not verify the second.

Thus, a law of error does not impose, in a formal way, any value for the measured quantity.

Bertrand had demonstrated the inverse proposition, in some sort: having been given an analytical method leading to a certain value u, it does not ordinarily correspond to it any law of error.

§ 1. — The Exponential Law and the Mean

We know that the exponential law has a special importance, such importance that, ordinarily, one studies it exclusively.

Or, if in the immense majority of cases, one law of error does not impose, in an absolute fashion, any value for the measured quantity, there is a specific exception when it is a matter of the exponential law.

The exponential law requires that one adopt for the measured quantity the arithmetic mean of the measurements, that is to say the quantity $(u_1 + u_2 + \ldots + u_n)/n$.

The arithmetic mean satisfies two conditions which were imposed for which a value would be necessarily preferred; it imposes thus in complete logic.

We are thus led to some very satisfying results:

At first, there exists a law of error of which the importance is absolutely preponderant.

Then, that law imposes in a logical fashion a certain value for the measured quantity.

In the third place, that value is obtained by a very simple process: just take the arithmetic mean of the measurements.

A fourth circumstance arises from the fact that the error made on the result while using the arithmetic mean, is itself the exponential law. The error committed on the result varies, in general, by the reciprocal of the square root of the number of observations.

§ 2. — Postulate of Gauss

It is to Gauss that one must go to find the law of errors. He took for a basis for the demonstration of the law a principle that one calls the postulate of Gauss.

Gauss, it must be said, had never considered his principle as a postulate. This scholar of genial and universal spirit would have refrained from presenting as a necessary truth an *a priori* assertion, very interesting notwithstanding, but not offering the characteristics of a postulate.

Here is the statement of principle of Gauss:

The arithmetic mean of the measurements is the most probable value of the quantity measured.

Departing from this principle, Gauss would demonstrate that there could only exist one law of error, the exponential law that one often called, for that reason, the law of Gauss.

Obeying an exaggerated faith, some have considered the principle of Gauss as evident.

It corresponds to an idea of necessary simplicity on which a mind as fine as that of Gauss could not delude himself.

That the arithmetic mean is a good value, no one disagrees, but that something of such precision as the arithmetic mean of measurements is something as precise as the most probable value of the measured quantity is inadmissible as principle. The assertion would not have value if one knew only his conclusions were verified by the experience and explicable by one hypothesis based on the nature similar to randomness.

To say that by instinct one takes the arithmetic mean proves nothing when it is a matter of obtaining something of infinite precision as an analytic formula.

To say that obeying the same instinct one usually eliminates guesswork, the values which diverge too much from the mean prove nothing more to invalidate the same analytic formulas.

The idea of Gauss, taken for what it is worth, was not very much less interesting and worthy of his great mind.

From Bertrand, the principle of Gauss must be modified; instead of pretending that the arithmetic mean of measurements is the most probable value of the measured quantity, one must say: "The arithmetic mean of the measurements is the mean value of the measured quantity."

That principle is evidently more plausible however insufficient to serve as a unique basis for a theory.

H. Poincaré had done interesting research on the postulate of Gauss modified by Bertrand. For that study, I can only return to his work on the calculation of probabilities.

Some have wanted to demonstrate the postulate of Gauss; the hypotheses on which they base themselves are much less plausible than the postulate itself.

That which is necessary to fully understand, is that the exponential law of errors requires that one adopt for the quantity measured the arithmetic mean of the measurements. If one assumes that the arithmetic mean of measures is the most probable value of the measured quantity, one assumes, for that same reason, the exponential law.

As soon as one adopts, *a priori*, for the value of the measured quantity the arithmetic mean of the measurements, one must obey three principles, of different claims, or simply a practical rule.

The principle of the arithmetic mean that one would be able to qualify from *absolute* would assume that the arithmetic mean is absolutely required, that it is at the time the mean value, probable and more probable than the measured quantity and that the equidistant values of that mean correspond to the same probability for the measured quantity.

The principle of Gauss assumes only that the arithmetic mean of measurements is the most probable value of the measured quantity.

The Bertrand-Poincaré principle assumes that the arithmetic mean of measurements is the mean value of the measured quantity.

Beyond these *principles*, the *rule* of the arithmetic mean is to adopt the arithmetic mean of the measurements in justifying only that choice for reasons of simplicity — a topic that we will revisit.

There is no evidence, *a priori*, that there exists a law of error satisfying the "absolute" principle. Alone, the exponential law verifies the principle of Gauss, as that law satisfies the conditions imposed by the absolute principle, this last principle and those of Gauss are, *a posteriori*, equivalent.

The Bertrand-Poincaré principle is finally equivalent to the absolute principle.

Thus, in reality, the three principles are only one.

Why would one want that the arithmetic mean of measurements possess *a priori* a precise and remarkable property? Because it is simple? The argument is of a great weight when it is a matter of a "rule," it is absolutely insufficient when it is a matter of a "principle."

§ 3. — Determination of the Precision

Where the law of error is exponential, one must adopt for the measured quantity the arithmetic mean of the measurements, that is to say the value $u=(u_1+u_2+\ldots+u_n)/n$. This quantity varies little from the exact value.

The more or less concordance of the values observed u_1, u_2 … u_n permits the forming of an idea of more or less the precision of the measurements.

From the value u one successively subtracts each of the values u_1. u_2,… u_n; one obtains thus the quantities $(u - u_1)$,$(u - u_2)$,…$(u - u_n)$ which are not exactly the committed errors(since u is not the exact value), but which differs little from it.

One adds these values $(u - u_1)$, … while considering them all as positive (that is to say that one adds their absolute values) and one divides this sum by the number n of observations.

The result is called the *mean error.*

In multiplying the mean error by 0.84, one obtains the *probable error*, that is to say the error which has as much chance of being or of not being exceeded.

The law being normal, the knowledge of the probable error allows, as we have seen, to calculate the probabilities of all the errors.

One often determines the precision or (what amounts to the same) the probable error by another process making use of, not more of these absolute values of the differences $(u - u_1)$,$(u - u_2)$, … but the squares of these differences.

One adds the squares of the differences, that is to say the squares $(u - u_1)^2$, $(u - u_2)^2$, … $(u - u_n)^2$, one divides by the number n observations and one extracts the square root of the result; one obtains thus that which one calls *quadratic mean error.*

In multiplying the quadratic mean error by 0.67, one obtains the probable error.

When the number of observations is very large, the use of the method of squares is preferable; Gauss has demonstrated that 100 observations with the use of that method permits determining the probable error with the same exactitude as 114 observations with the use of the first process.

Where the observations are less numerous, the use of the base method on the squares of the errors must be thrown out because it is too dangerous.

If a caprice of randomness produces a huge error, the square of that error will be exaggeratedly large and will be able to have a huge influence on the result which will be very different from the value sought.

Any formula containing squares is thus inapplicable when the number of observations is small. In the long run, the increasing number of observations, from compensation, happens and the process that was used that was condemnable finishes by becoming the better.

One could employ some other methods for the determination of the probable error, they are generally more complicated than the preceding and require that the observations be much more numerous.

One very simple and very natural method consists of determining the probable error from its definition. One would classify the positive errors by order of size and would adopt for the probable error that which occupies the middle of the ranking.

This process is very nearly precise, 250 observations would be necessary to obtain the probable error with the same exactitude as by the use of only 100 observations and from the method of the sum of the squares.

From the purely logical point of view, any method of determination of the probable error only imposes an absolute manner.

§ 4. — Law of Ordinary Error

Errors nearly always follow the exponential law or normal law or law of Gauss, because the hypothesis of infinitesimal errors is nearly always true. It is meanwhile useful, for the generality of our study, to envisage the case of a law of ordinary error.

Thus we assume the law of ordinary error, but given and unique for all the observations. We know that, ordinarily, one cannot deduce the observed values u_1, u_2, … u_n a value which imposes in full rigor for the quantity measured.

For lack of something better one uses then the *rule of the arithmetic mean* which consists of adopting the arithmetic mean of the measurements, that is to say $u = (u_1 + u_2 + \ldots + u_n)/n$.

That choice presents some advantages: the process is simple and, if the observations are numerous, the error committed on the result follows the law of Gauss.

The error of the result varies, in general, by the reciprocal of the square root of the number of observations, as if each error individually follows the exponential law.

That conclusion could seem paradoxical; in assuming it, one would approach the exact value indefinitely while augmenting the number of observations, that which appears physically unrealizable.

That it is physically unrealizable, is the hypothesis of the absolute nullity of the systematic errors, the conclusion is only the logical consequence of the hypothesis.

The decreasing of the error committed on the result while admitting the nullity of the systematic errors is very slow, it proceeds following the square root of the number of observations; it is necessary to centuplicate this last number in order to obtain ten times more precision.

It is almost clear that the error committed on the result must follow the law of Gauss when the observations are very numerous.

In effect, the sum of the errors follows necessarily the law of large numbers; each error is, relative to the sum, an infinitesimal error.

The sum of the errors follows the exponential law.

The arithmetic mean of errors is, by definition, the sum of the errors divided by the number of them, it must follow thus the law of Gauss.

The error committed on the result, while employing the rule of the arithmetic mean is the arithmetic mean of the errors: it is thus the law of Gauss.

The error committed on the result, while employing the rule of the arithmetic mean, follows the law of Gauss when the observations are numerous; it follows *rigorously* that somewhat little law which is the number of the observations when, for each of them, the law of error is exponential.

The law of Gauss or exponential law thus possesses that excessively remarkable property of self-reproduction by composition; if two or more quantities (for example of errors) follow the law of Gauss, their sum and their arithmetic mean follows rigorously that same law. The importance of that result has been put into evidence by M. d'Ocagne.

One must be wary of believing that the probable error committed on the result, when the law of error is ordinary, is equal to the probable error of an observation divided by the square root of the number of observations, that is only true if the law of error is exponential. In the case of a law of ordinary error it is the mean quadratic error of the result which is equal to the mean quadratic error of an observation divided by the square root of the number of observations. The mean quadratic error has been defined precisely, it is the square root of the mean of the squares of the errors.

§ 5. — THE MEDIAN VALUE

Instead of using the rule of the arithmetic mean one could make use of the *rule of the median value*; it consists of ranking the obtained values in order of size and of adopting that which occupies the middle of the ranking.

If the number of observations is even, one adopts the arithmetic mean between the two values which occupy the middle of the ranking.

The rule of median value presents some advantages analogous to the rule of the arithmetic mean; its use is simple and, if the observations are very numerous, the error committed on the result follows the law of Gauss.

Laplace has studied the method of the median value which he named *method of position*; he demonstrated that, following the form of the law of error, there was advantage to using the rule of the arithmetic mean or the rule of the median value and that an appropriate combination of the two rules measured better, in general, than the exclusive use of one of the methods. He encountered an exceptional case, that of the law of exponential error; in that last case one must use exclusively the arithmetic mean and 100 observations by following this rule valuing 125 observations in following the rule of median value.

That result should not surprise us, we know that the arithmetic mean is required when the law of error is exponential.

We know also that in general, any value not requiring an absolute method, a combination of the two preceding rules would be able to lead to a value which, *in general*, will be preferable to the others; a different process, maybe, leads to a value which will be again preferable to certain points of view, no value can be required.

Practically, one uses by preference the rule of the arithmetic mean and one does fine, but that rule is not required outside of the case of the exponential law. It is likewise very curious that it is not always required where there are only two measurements made; the arithmetic mean could give a *minimum* probability to the fact observed, that is to say to the production of two equal errors and of opposite signs. One meanwhile adopts the mean by reason of symmetry, one adopts it because it is absolutely necessary to adopt a value.

§ 6. — DISSIMILAR OBSERVATIONS

Where the observations are not equally precise, it is no longer rational to adopt the arithmetic mean of the measurements, the most precise observations come to have more influence on the result.

One then makes use of the rule of the *mean by weight* [weighted average], it consists of taking the mean only after having multiplied each measurement by a coefficient representing the degree of exactitude which one knows of it or that one assumes of it.

That coefficient is the *weight* of the observation.

The expression of mean by weight, from a mechanical origin, is easily interpreted.

Let us imagine a straight line, rigid and without mass. Starting from a point on that line, let us take on that and in the same sense of equal lengths to different measurements u_1, u_2, ... u_n. From each of those points thus obtained, let us suspend a weight proportional to the points of the corresponding observation. The rule of the mean of the weights consists of adopting for a value of the measured quantity that which corresponds to the center of gravity of the different weights.

That which renders the analogy even more understandable is that the weight of the result is equal to the sum of the weights of the observations.

When, at each observation, the law of error is exponential (with a coefficient of precision or a probable error variable from one observation to another), the rule of the mean of weights is required in an absolute fashion.

Beyond this case, it is only necessary to consider the rule as a simple and convenient process, presenting some analogous advantages to those of the rule of the arithmetic mean to which it reduces itself even when the observations are identical.

§ 7. — Discordant Observations

If some measurements deviate too much from the mean of the others, what position must one take in their respect? Must one reject without other forms of process and adopt a new mean? Must one keep them at the same dignity as the others? Should one keep them while diminishing their weight? And in what proportion?

There enters so much arbitrariness in this question that one cannot resolve it in a satisfactory manner. Astronomers, who the subject interests in particular, each have their opinion and their prescriptions are very different.

"... By what right to reject a measurement who diverges much from the mean?" say the partisans of maintaining integrity. "Since the observation has been made in the same conditions as the others, with the same care, with the same precautions, you are not authorized to suppress that measurement. You must only consider the effects of randomness and permit yourself to correct the caprices; you distort the result.

"... It is precisely because we don't want to consider only the effects of randomness that we reject that measurement," say the others, "we do not place into doubt the sincerity of the observer, he is assumed to have taken the same care as ordinarily, but he is fallible; is it not more logical to assume that he has had an involuntary distraction than to believe to have arrived from an excessively unusual randomness? A distraction could produce an error which is neither systematic nor fortuitous, which is, in some way, exceptional; one must reject an observation very probably tainted by such an error, it would unjustly make the mean deviate.

"—You want to condemn that measure because it would make to bend the balance to one side, but, if you suppress it, the balance will be bent to the other; it is necessary to consider this observation; for mediocre such that it would be, it would be worth something; by rejecting it you probably suppress an error to the right in order to introduce one to the left."

One would be able to discourse for a long time on the subject without arriving at agreement; no solution is satisfying.

Legendre was partial to the successive approximations: after having taken the arithmetic mean of measures, one suppresses, "by guesswork," the measurements which deviate much from its mean. One adopts then the arithmetic mean of the remaining measurements.

That process is imprecise and arbitrary, like the others besides.

Svanberg adopted as first approximation the arithmetic mean and he determined the weight of each observation from after the relative difference to the mean; he applied then the rule of the mean by weight.

Laplace had demonstrated that the second approximation did not validate the first: 120 observations with the method of Svanberg validated only 100 observations with the arithmetic mean. But Laplace assumed little errors and numerous measurements; this is not the case on which we occupy ourselves, we study, instead, the case where the observations are few and where certain errors have an exceptional size. The method of Svanberg, for which the theoretical superiority is not demonstrated, has the inconvenience of being complicated and, therefore, difficult to apply.

Faye, the famous astronomer, was partial to the maintenance of the integrity of the measures, except, necessarily, where the observer believed not to have observed with the same ordinary care. In this last case, it is necessary to diminish the weight of the observation.

Asaph Hall, astronomer equally famous, to whom we owe the discovery of the satellites of Mars, was partial to maintaining the integrity of measurements and saw a great inconvenience in the suppression of those which appeared discordant: if one eliminates the discordant measures, those that remain will be exaggeratedly concordant; then, in the future, one will make themselves an idea much too optimistic on the precision of the result. In such matters, the optimism is more dangerous than the converse excess.

Certain astronomers have proposed some "criteria of rejection" for the doubtful observations, others have recommended some methods for which the weights of the discordant observations will be diminished. These rules could be useful, but they behave very arbitrarily.

The arithmetic mean of the measures appear more affected by the exceptional errors than the median value. Two exceptional values, one by mistake, the other by excess, doesn't change the median value, it could influence very sensitively on the mean.

A mean error or a very large error effects the median value equally. The arithmetic mean, on the other hand, nearly indifferent to the mean error, is absolutely deflected by the large error.

The law of Gauss, or exponential law, grants only one very weak probability to large errors. Where they produce many too frequently, it is necessary to adopt another law, and one falls into arbitrariness.

One could try a law formed by the combination of two, of several, or from an infinite number of laws of Gauss, each of them corresponding to a coefficient of different precision.

Simon Newcomb used this process to explain the errors which were found in paths of Mercury around the Sun. We are led, by an absolutely necessary fashion, to these laws in the study of the statistics of speculation.

These laws of errors are analogous to the law of Gauss, but they give a larger frequency to very small errors and to very large errors, to the costs of these mean errors of which the frequency is diminished.

The curves representing these laws of error are analogous to the curves of Gauss, but they are higher at their peak and in the elongated tails, less high in the middle parts.

§ 8. — Quantities Following the Normal Law

Certain quantities, in their differences relative to the mean, follow the normal law of probability as if randomness, alone, produced these differences.

We have noted for a very long time that it was thus for the differences in the height of conscripts [people of the same age]; these differences follow nearly the exponential law as if they were due uniquely to chance.

One measures, for example, the height of 100,000 conscripts and takes the mean; it is, I'm assuming, 1.65 m. The difference between the height of a conscript and the mean, 1.65 m, is an écart.

One notes at first, by the statistics, that there are as many positive écarts as negative écarts, that is to say that there are as many conscripts whose height is more than 1.65 m as conscripts whose height is less than 1.65 m. It is necessary to keep from believing that this result is evident; it would be absolutely inaccurate if one considers the weights of the conscripts instead of their heights; there are many more than half of the conscripts whose weight is less than the mean of the weights.

One notes then by the statistics that the small écarts are more frequent than the large and that the positive écarts have similar frequency as the negative écarts of the same amplitude. There are as many conscripts whose height is greater than 1.68 m (écart greater by +3) as conscripts whose height is less than 1.62 m (écart less than -3).

In studying the statistics more closely, one notes that the écarts follow the exponential law, or the law of large numbers, as the écarts in the game of heads or tails, as the errors in the measurements.

The probable écart is that which is exceeded one time in two; it will be, for example, 5 cm. There will be thus around 25,000 conscripts whose height will be less than to 1.60 m; 25,000 for whom the height will be comprised between 1.60 m and 1.65 m; 25,000 whose height will be comprised between 1.65 m and 1.70 m, and 25,000 whose height will be greater than 1.70 m.

Since it is a matter of the exponential law or of large numbers, the knowledge of the probable écart drives the knowledge of the probabilities of all the other écarts. We know, for example, that there is a 0.177 probability for which the écart twice that of the probable écart will be exceeded; we could thus predict that of the 100,000 conscripts considered, there will be 8,800 of them whose height will be greater than 1.75 m, and 8,800 whose height will be less than 1.55 m.

Relative to the height of the conscripts, there are thus two quantities to consider; the first, very important from the point of view of the evolution of the race, is the general mean of the height. The second measures the écarts on either side of that mean value; this is the probable écart.

One considers as an index of the purity of the race the fact that the height of the adults follows the exponential law or of large numbers. If the French, instead of being constituted by a homogeneous race, was one half dwarfs and one half giants, the mean of the heights would have to be the same; but the écarts, relative to this mean, would no longer have the same frequency. The little écarts, instead of forming the great majority, would only be more exceptional; there would no longer be a unique height having a maximum frequency, but two corresponding heights, one for the dwarfs, the other for the giants.

— Chapter XXIII —

Curves of Frequency

Let's consider a quantity susceptible of taking a certain number of values: it could be, for example, the height of the conscripts, the volume of the brain, the daily change of the price of the bond, the wealth of individuals belonging to a nation, etc.

By observation, we could determine the frequency of the different values of the considered quantity, we thus form the statistic of it.

The obtained values, faithfully transcribed from the results of the observation, have the disadvantage of not giving a very clear and very accurate general idea of the results; also there is often interest in constructing the curve representative of the results, called the *curve of frequency*.

One takes for the horizontal axis the different values of the quantity and the corresponding frequencies for the vertical axis. One often chooses for the origin the mean value of the considered quantity.

For example, if it is a matter of the height of French conscripts, the écarts follow the normal law, the curve of frequency will be a curve of randomness having the shape of a sort of bell, a curve that one even calls curve of probability, normal curve, curve of Gauss, curve of Quetelet, or binomial curve.

If it is a matter of the height of French women, one would obtain an analogous curve, maybe a little more or a little less flared, a curve having the form of a sort of bell.

The result will be different if one represents graphically the écarts of the heights of the adults of both genders: the general mean of the heights, that one takes for the origin, would retain the middle between the mean of the heights of men and the mean of the heights of women.

The difference between these means is not sufficient for which the curve of frequency would present an undulation (like that of figure 3), but it is sufficient so that the curve differs markedly from the normal curve and would be much flatter in the middle.

Figure 3 shows what would be the representative curve of the heights of the male adults in a country where there were two pure races, equally numerous and with very different mean heights; one has taken for the horizontal axis the écarts relative to the mean general height, and for the vertical axis the corresponding frequencies. It is not the general mean that corresponds to the highest frequency.

It is very often interesting to compare some curves of frequency relative to some analogous quantities, for example the heights of conscripts in different countries. The comparison by the curves of frequency themselves is sometimes rendered rather troublesome by the entanglement of these curves. The graphic then loses one of its great qualities, which is the clarity and the readability and some other representation could be preferred.

One can see in the four graphs that follow a different mode of representation of the ordinary process. These tables are taken from a memoire of Dr. Gustave Le Bon, a memoire which was crowned by the Institute and which has for a title: *Anatomic and Mathematical Research on the Laws of Variations of the Volume of the Brain and on their Relations with Intelligence.*

It deals with the studying the variation of the volume of the brain according to the gender, according to brain activity, according to the period, according to the country. The use of the traced curves of frequency according to the usual process, by taking for the horizontal axis the different values of volume of the brain and for the vertical axis the corresponding frequency has presented two inconveniences; the curves would be entangled and, for the other part, the extreme values would have been very badly figured.

On the other hand, the four tables that we have reproduced are of a beautiful clarity and one can deduce from them, by a simple look, some very interesting results. Let us consider, for example, the upper line of the first graphic, that which is relative to modern Parisians of the masculine sex; as the ordinate 1,650 corresponds to the abscissa 800; we must thus conclude that of 1,000 Parisians, 800 have a brain size less than 1,650 cubic centimeters. At the ordinate 1,430 corresponds the abscissa 150, we must thus conclude that, in 1,000 Parisians, 150 have a brain volume less than 1,430 cubic centimeters.

The ordinate represents the cubic centimeters, the corresponding abscissa represents the total of the individuals who have a brain capacity less to that number of cubic centimeters.

Let's assume, for example, that we want to know the number of male Parisians who have a brain capacity comprised between 1,500 and 1,600 cubic centimeters. The ordinate 1,600 corresponds to the abscissa 720. There are therefore 720 male Parisians in 1,000 for whom the brain capacity is less than 1,600 cubic centimeters. One sees that likewise there are 240 of them out of 1,000 for whom the brain capacity is less than 1,500 cubic centimeters. There are thus 720 - 240 = 480 out of 1,000 for who the brain volume is comprised between 1,500 and 1,600 cubic centimeters.

From the first table we could, immediately, pull some interesting conclusions. The volume of the brain of the male Parisian is very superior to the volume of the brain of the female Parisian (same relationship to their respective weights) and the écarts on either side of the means are much greater for the males. The volume of the brain of the female Parisian is only slightly larger, on average, to the volume of the brain of the males from inferior races, but the écarts are much greater.

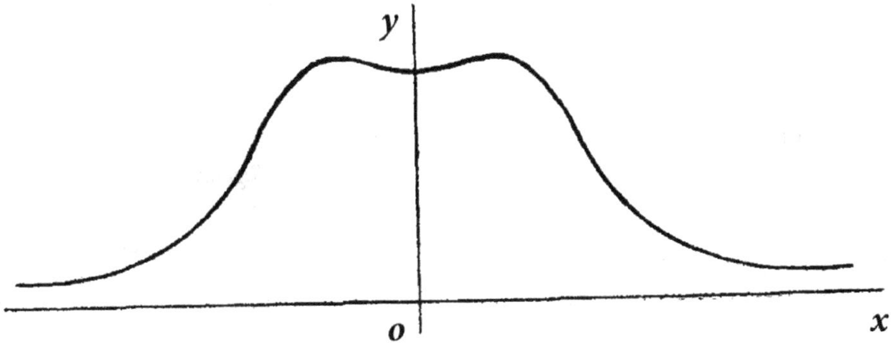

Figure 3

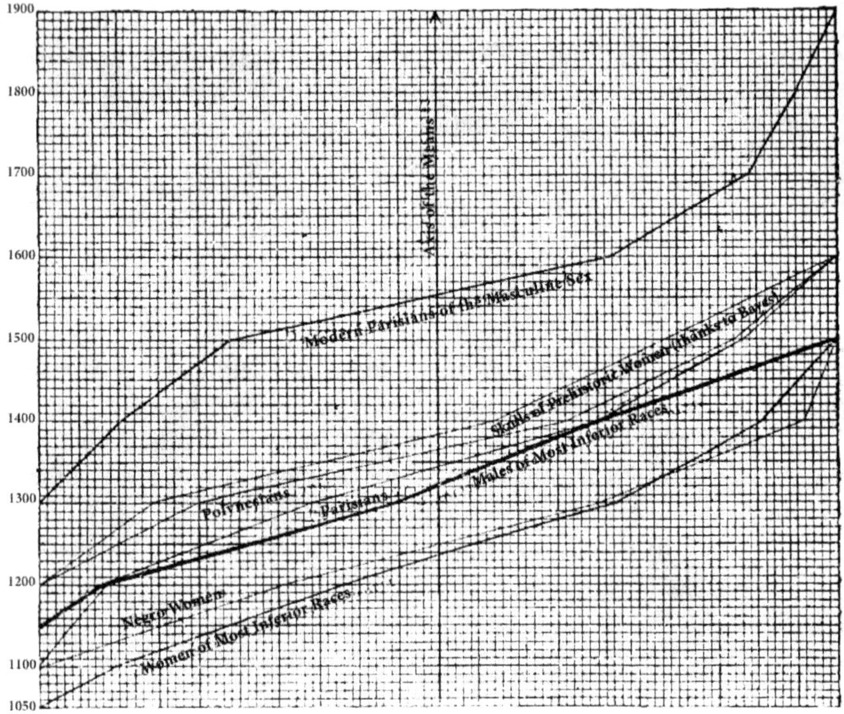

Table I is taken from a memoire of Dr. Gustave Le Bon, having for a title: *Anatomic and Mathematical Research on the Laws of Variations of the Volume of the Brain and on their Relations with Intelligence.* These curves show evidence of the variation of the volume of the brain according to sex and race. The abscissa indicates the number of individuals (out of 1,000) who have a brain capacity less than a value expressed by the corresponding ordinate. For example, on the curve relative to modern Parisians, an abscissa 400 corresponds to the ordinate 1,530. There are therefore 400 Parisians out of 1,000 whose brain capacity is less than 1,530 cubic centimeters.

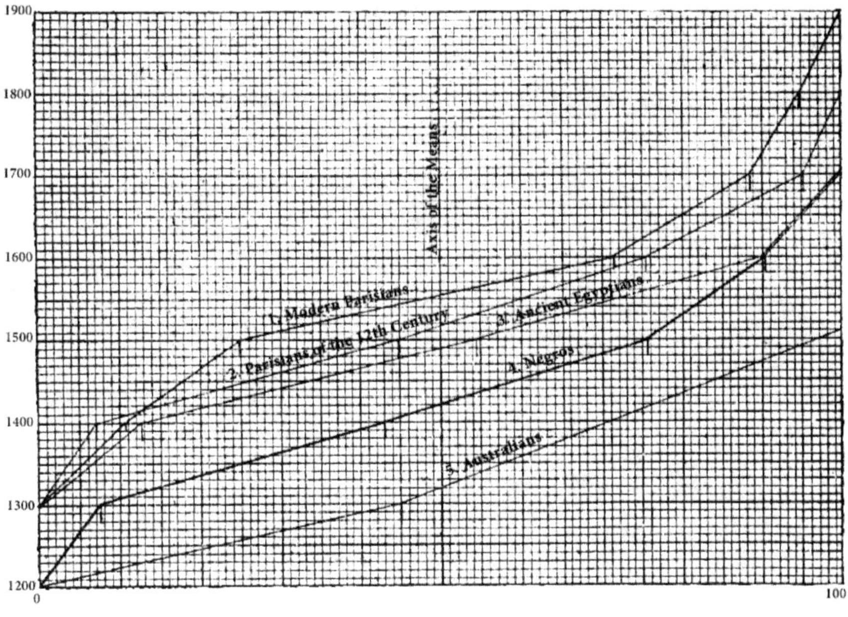

Table II taken from the previously cited memoire by Dr. Gustave Le Bon. These curves show the progressive development of volume of the brain in the human races. The mean of the volume has increased through time as the deviations from either side of the mean. If one wanted to know, for example, the number of male Parisians whose brain volume is comprised between 1,600 and 1,800 cubic centimeters, simply note that at the ordinates 1,800 and 1,600 correspond the abscissa 950 and 710. There are thus 950 - 710 = 240 male Parisians out of 1,000 whose brain size is comprised between the given limits.

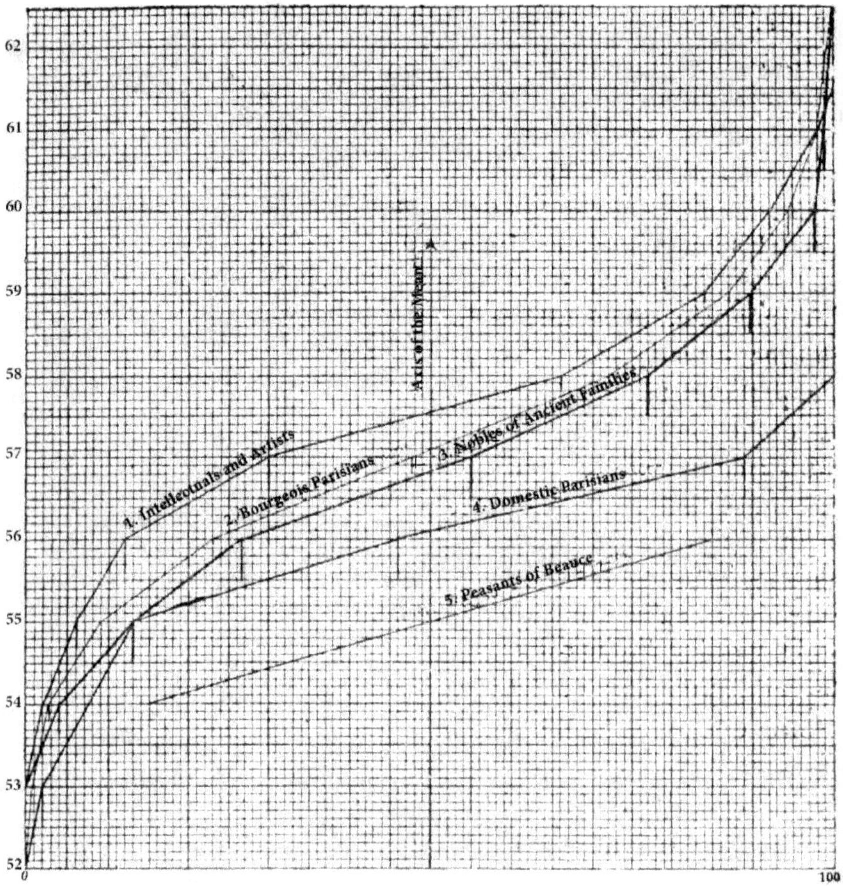

Table III taken from the previously cited memoir of Dr. Gustave Le Bon. These curves show evidence of the relationship which exists between the circumference of the head and intellectual activity.

The second table indicates the variation of the brain volume through time; it seems that civilization has had the effect of augmenting the mean of the volume in the same way as the écarts relative to that mean.

The third table shows the influence of the intellectual culture on the development of the brain; the values written in the left scale (ordinates) represent, in centimeters, the circumferences of the heads.

One gives the name of biometry to the science which studies the statistics relative to the measurements of organic beings. That science is especially beholden to Quetelet who was, one could say, the creator.

§ 1. — Analytic Representation of the Curves of Frequency

Let's return to the ordinary mode of representation, one takes as the abscissas the different values of the quantity considered and as the ordinates the corresponding frequencies; one obtains thus a curve which, *a priori*, is whatever.

When the curve is traced, one can look to see if it doesn't present the resemblance with some known curve for which the equation is simple and contains only a small number of parameters.

If the resemblance is adequate, the equation of the curve gives a sort of analytic representation of the law of frequency.

But that analytic representation can only have some other pretention to give a general idea of the approximate values of the frequency, it must not have the pretention of expressing realistically the true law of frequency when that law exists.

Two very different formulas from the analytic point of view, which would have been the translation of two phenomenon completely different in their essence, could lead to some very nearly equivalent numerical results between the rather extended limits of the variable, especially where these formulas contain several parameters. It is not necessary to believe that one necessarily knows the formula which in reality regulates a phenomenon because that formula leads to some numeric values neighboring those which give the observation;

the true formula is maybe analytically very different than those that we consider. Practically, it matters little and it is why the analytic representation of certain phenomenon, of mortality, for example, is so useful.

§ 2. — The Mean

In many cases, one is obliged to represent by a single number a set of values; one then adopts the ordinary mean.

One also adopts sometimes the value which occupies the middle in the ranking by order of size of the considered values.

Let's assume that one adopts the mean.

The mean, a unique value, cannot have the pretention of representing the values that it summarizes.

If I find myself in the obligation of stating one rather naïve truth it is that one often attaches to the knowledge of mean values more importance than it deserves.

The knowledge of the mean value is more or less useful following the nature of the considered quantity, following the mode of grouping more or less symmetric for the different values around that mean, following the amplitude of the deviations relative to that mean.

The knowledge of the wealth in a country where there are only millionaires and impoverished people does not present the same interest as the knowledge of that mean for a country where the fortunes are distributed more uniformly. The mean, unique value, cannot exactly represent a multitude of other values.

One number cannot represent several others if one takes a particular point of view. The knowledge of the product of sides of a rectangle equates to the knowledge of the sides of that rectangle if it is uniquely a matter of the area of it; the two knowledges are no longer equivalent if it is a matter of a rectangular terrain on which one must build an edifice.

It is very curious that one has sometime made a criticism about the notion of mean value not corresponding to a possible value of the quantity considered while that the same criticism was not directed in respect to the analogous notion of the center of gravity.

The notions of the mean value and of the center of gravity are identical; the center of gravity doesn't have any more real existence than the mean value, sometimes it falls outside of the considered body in a point empty of matter. The mean value does not define more than the values it summarizes than the center of gravity defines the system to which it belongs.

§ 3. — The Dispersion

The mean value gives a first general idea of the values of a quantity; a second value will give a general idea of the differences [*écart*] relative to that mean.

The écart for a determined value is the difference between that value and the mean.

The numbers that give a general idea of the amplitude of the écarts are called *coefficient of dispersion.* One usually adopts for this coefficient the quadratic mean of the écarts or the quadratic écart.

The quadratic mean of a quantity is the square root of the mean value of squares of that quantity.

To obtain the coefficient of dispersion, one thus calculates the mean value, one subtracts that from each of the obtained values, one thus obtains the differences. One takes the arithmetic mean of the squares of these differences and one extracts the square root of the result.

One equally uses for coefficient of dispersion the mean value of the positive differences or mean difference.

We have already considered these same quadratic differences and means in the particular case of the errors of observation. In this particular case, the differences follow the normal law, the relationship of the quadratic differences and mean is a fixed number, 1.25. It is not the same in the general case.

It is very evident that a coefficient of dispersion, as a unique value, cannot represent by it alone all the differences that it summarizes, it can only give a general idea of the amplitude of these differences.

One sometimes makes use of a third coefficient called *coefficient of dissymmetry*, destined to give an idea of the dissymmetry of the distribution of the differences from either side of the mean or of the dissymmetry of the curve of frequency.

The mean of the cubes of the differences gives an idea of the dissymmetry of the curve. If, in effect, the curve was symmetrical, each positive difference would correspond to a negative difference of the same amplitude and the mean value of the cube of the difference would be null. In order to obtain a coefficient, one would divide, for example, the cube root of the mean of the cubes of the differences by the quadratic difference or by the mean difference.

§ 4. — Covariations

When two quantities depend one on another in such a way that every value of the one imposes a value on the other, one says that these quantities are functions of one another.

Two quantities could depend on one another without their relationship being absolutely of a functional nature.

The length of the arm of an individual is dependent on his height, but if one knows the height of an individual one cannot deduce from it exactly the length of his arms.

The length of the arms of an individual depend on his height and on a bunch of other factors that we cannot analyze. What we desire, is simply to make for us a general idea more or less of the relationship which exists between the height of one individual and the length of his arms.

The memoire of Dr. Gustave Le Bon, who has already been useful to us, did not offer to establish a relation of the functional nature between the volume of the brain and intelligence; it has never been in the thought of its author to pretend that one individual for whom the brain is small is necessarily of little intelligence nor that an individual for whom the brain is very developed is necessarily a genius. He only wished to establish that there exists a general relationship between the volume of the brain and intelligence.

The relationship which exists between two quantities generally depends on an array of factors; our pretension must not be to measure that relationship, it must be limited to obtaining a general idea more or less of the mutual dependence of the considered quantities.

Toward this end, the examination of the curves of frequency is very useful; the fourth table taken from the previously cited memoir of Dr. Gustav Le Bon shows the weak influence of the height on the weight of the brain.

The scale on the left, in the margin, indicates the weights of the brain of 900 to 1,700 grams.

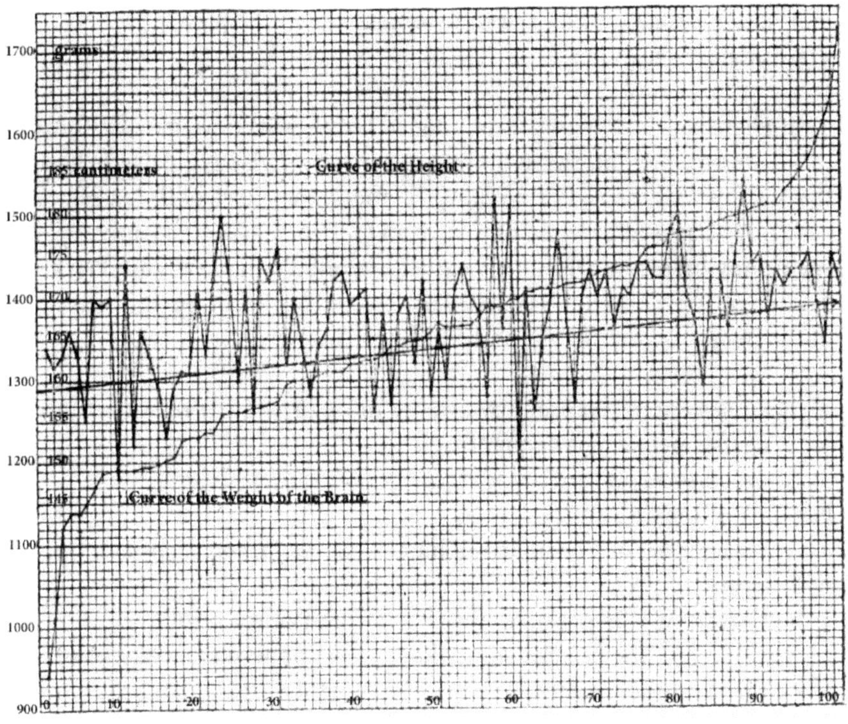

Table IV taken from the previously cited memoir of Dr. Gustave Le Bon. These curves show evidence of a small influence of height on the weight of the brain. At the abscissa 40, for example, corresponds on the curve of weights of the brain to the ordinate 1,320 (counted on the scale which is outside of the grid, in the margin) and the ordinate 170 on the meandering curve traced in a solid line (ordinate counted on the scale which is inside of the grid). It is necessary to conclude that in 1,000 subjects there are 400 whose brain weighs less than 1,320 grams and that the subjects whose brain weighs 1,320 grams have a mean height of 1.70 m.

The left scale, in the grid, indicates heights of 1.45 m to 1.85 m. Let's consider for example the number 1,200, in the scale in the margin, it corresponds to the abscissa 15 on the curve representing the weights of the brain; we must thus conclude, as previously, that, in 100 individuals, there are 15 of them whose brains weigh less than 1,200 grams.

At that abscissa 15, corresponding on the irregular curve traced in solid line an ordinate (counted on the scale inside the grid) equal to 158; we conclude that the individuals whose brain weighs 1,200 grams have a mean height of 1.58 m.

If there exists a strong dependence between the weights of the brain and the height, the solid line curve would rise regularly from the left towards the right. In reality, one sees, the curve of the sizes is very irregular and likewise of the mean (the mean is represented by the right which cuts the figure nearly horizontally in the middle) it is very slightly ascending. It is necessary to conclude that the influence of the height on the weight of the brain is very weak.

One can try to represent by a value more or less the relationship which exists between two quantities.

I insist on this point, it is not a matter of measuring the relationship, but only of making an idea.

The value that one ordinarily adopts is the *coefficient of correlation* of Pearson.

Let's assume, for example, that it is a question of setting an idea of more or less the relationship which exists between the height of the father and the height of the son. Let's further assume, in order to simplify, that the mean of the height of the fathers' is 1.65^{m}, the same as the mean of the height of the sons.

The écart is ordinarily the difference between the height itself and the mean; if, for example, the height of an individual is 1.70^{m}, the écart is 5, if the height is 1.63^{m}, the écart is -2, etc.

If an écart for the height of the father makes necessary a *proportional écart* for the height of his son, it will have complete correlation. If for example, the écarts 3, 9, -6 for the fathers necessitates, respectively, the proportional écarts 2, 6, -4 for the sons, there would be absolute correlation, the coefficient of correlation would have its maximum value 1.

When these considered quantities are independent, the coefficient of correlation is null.

In the case of the dependence of the height of the sons in regard to the height of the father, the coefficient of correlation is a certain number included between zero and one.

The coefficient of correlation is negative when the considered quantities vary inversely. For example, the price of wheat and the quantity of wheat vary inversely; the greater the quantity the lower the price.

If, at each écart for the quantity of wheat relative to its mean, would correspond to a *proportional* écart for the price of the wheat, relative to its mean (the écarts having opposite signs), there would be an perfect negative correlation. The coefficient of correlation would have its minimum value, -1.

In summary, the coefficient of correlation varies between -1 and +1.

When the coefficient is equal to -1, there is absolute inverse correlation, when it is equal to +1, there is absolute direct correlation.

When the coefficient of correlation is positive, the quantities vary, in general, in the same sense; when it is negative, the quantities vary, in general, in a contrary sense.

The coefficient of correlation does not only reflect more or less the relationship between two quantities, it depends much on their proportionality more or less; this is why one names it sometimes *coefficient of covariation*. One could imagine two quantities having a very intimate dependence but vary, so to speak, antiproportionally, in such a way that of very large écarts of the one necessitates very small écarts of the other, the coefficient of correlation would have a small value although there is a relation of a functional nature. It is necessary to recall that this case is not very interesting from a practical point of view.

One substitutes sometimes for the coefficient of correlation another number called ratio of correlation for which the use presents some advantages.

One could propose to express by a value the relationship more or less which exists between two things which are not susceptible to a quantitative representation; for example the color of the eyes and the color of the hair.

We hasten to say that this value, called *coefficient of contingency*, does not nullify the pretention of measuring the relationship, it simply applies an idea to it.

If each nuance of the color of the eyes necessitated a nuance for the color of the hair, the coefficient of contingency would be around one. It would tend toward its maximum value one while at the same time as the number of nuances tended towards infinity.

If there wasn't any connection between the color of the eyes and those of the hair, the coefficient of contingency would be null.

The coefficient of contingency is thus always included between zero and one.

In concluding this rather brief study of the curves of frequency and the correlations, I again stress this point: the values obtained do not have the pretention of giving an absolute measure of the quantities that they summarize, they only give an idea. Nonetheless, their consideration presents a very great interest and a very real usefulness.

— CHAPTER XXIV —

THE SHOT AT THE TARGET

The theory of écarts in the shooting of a target is one of the most interesting of the calculations of probabilities; for a hundred years since it's been known, the experiment has always verified that it is a matter of the shot of the cannon or of a gun; its study leads to some very simple and meanwhile very curious results.

Let us assume that the gunman and the firearm of which he makes use have no systematic defects, that is to say that they don't have a constant tendency to pull sometimes to the right or to left of center of the target, sometime above or below.

Regardless of the position of the gunman and the quality of the firearm, the center of the target is not ordinarily attained, the projectile hits the target at a certain distance from the center. That distance is an écart and, since there is no systematic defect, that écart must be attributed to randomness.

The projectile has hit the target at a certain point (point of impact), the distance of this point to the center does not characterize the shooting; the simplest experiment shows, in effect, that the écarts in the vertical sense are, on average, greater than the écarts in the horizontal sense.

It is necessary to take into account and to consider for each projectile fired:

The vertical deviation or vertical écart, or again the écart in range or in elevation and the horizontal deviation, or horizontal écart, or again the écart of direction.

The écart was the distance of the point of impact to the center of the target, the écart in range is the projection of the écart on the vertical and the écart in direction is the projection of the écart on the horizontal.

In less precise terms, the écart in direction is the écart to the right or to the left, the écart in range is the écart toward the top or to the bottom.

The fundamental result of the theory of the shooting at a target is stated thus:

The écarts in range follow the normal law.

The écarts in direction follow the normal law.

More generally, if one projects the écarts on any line passing through the center of the target, the projections follow the normal law.

The vertical écarts, for example, follow the same law that the écarts in the game of heads or tails, in financial speculation, in ordinary measures, etc.

If one knows the probable vertical écarts, that is to say the vertical écart which has one chance in two of being exceeded (0.20^m over the level of the center of the target, for example, and 0.20^m under), one knows the probability for which another vertical écart would be exceeded. For example, one knows that the écart twice the probable écart has a probability 0.177 of being exceeded; there is thus 0.177 probability for which the projectile hit the target at more than 0.40^m over the plane of the center of the target or at more than 0.40m under.

There is a 0.043 probability for which the triple écart of the probable écart is exceeded; there is thus a 0.043 probability for which the projectile hits the target at more than 0.60m over the center of the target or at more than 0.60m under.

In general, knowing the probability for which a given vertical écart is passed, one knows the probability for which all other vertical écart will be exceeded.

If the gunman is skillful, the probable écart in height will be small and the other écarts in height will be proportional; if the gunman is unskillful, the probable écart in height will be greater and the others will be proportional.

The randomness will alone be the cause in the question studied, we find the same results as in the analogous case where randomness was alone.

The horizontal écarts follow the same law as the vertical écarts, but with another coefficient of precision, or, what amounts to the same thing, with another probable écart.

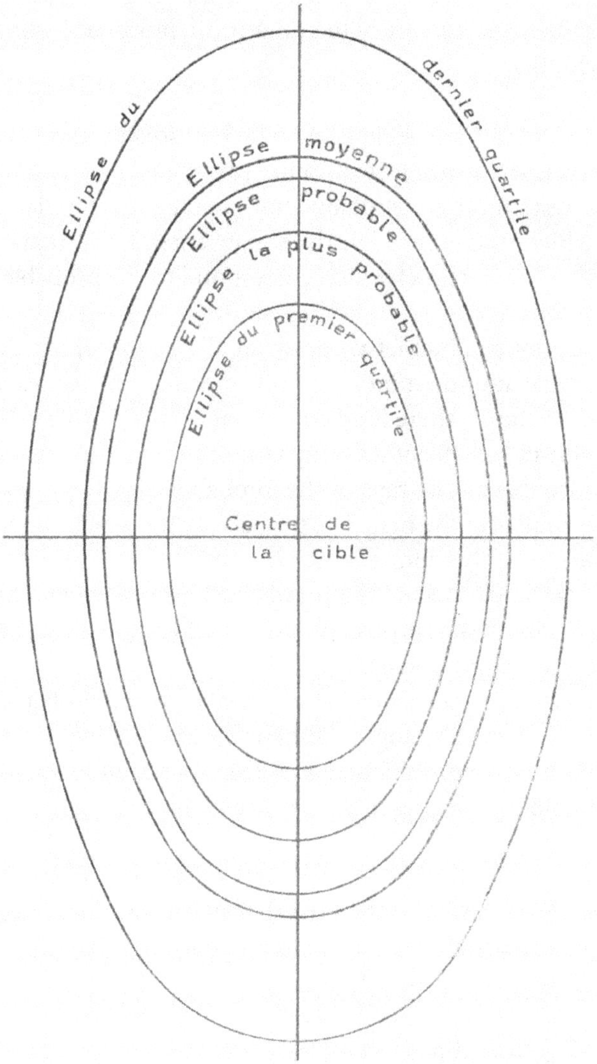
Ellipse du dernier quartile
Ellipse moyenne
Ellipse probable
Ellipse la plus probable
Ellipse du premier quartile
Centre de la cible

On average, the horizontal écarts are around two thirds of the vertical écarts, but the relationship varies according to the firearm and the gunman.

One shot is not uniquely characterized by the knowledge of the probable vertical écart and of the probable horizontal écart; to these two quantities it is necessary to add a third, however I won't bother myself with that.

On the straight horizontal line which passes through the center of the target, let us mark two points (on the right and on the left) which correspond to the probable horizontal écart. Similarly, on the vertical which passes through the same center, let us mark the two points (above and below) which correspond to the probable vertical écart.

By these four points let us draw an ellipse for which the major axis is vertical and the minor axis is horizontal; this ellipse is called the probable ellipse; there is an equal chance that a projectile hits the target inside or outside of this ellipse.

Let us double all the rays of this probable ellipse, we obtain thus an ellipse similar to the first. The probability for which the projectile passes outside of this ellipse is 0.177.

Let us triple the rays of the probable ellipse, we obtain an ellipse which is similar to it. The probability for which the projectile passes outside of this ellipse is 0.043.

If one considers an ellipse at all similar to the probable ellipse, that is to say formed by augmenting or by diminishing its rays in the same ratio, all the points situated on the contour of that ellipse have equal probability of being hit by the projectile.

Knowing the probability for which the projectile passes outside of a certain ellipse similar to the probable ellipse, one can calculate the probability for which the projectile hits outside of all other equally similar ellipse to the probable ellipse.

That property is strongly curious, we see that, geometrically, the shot is characterized by the probable ellipse.

The figure represents five principal ellipses; the first that one meets in starting from the center is the ellipse of the first quartile, there is one chance in four that the projectile falls inside of that ellipse. These rays are two thirds of those of the probable ellipse.

The second ellipse is the most probable ellipse, that which has a greater chance of being attained by the projectile. The probability that an ellipse (theoretically an infinitely thin elliptical ring) will be achieved, is even greater, firstly, as that ellipse is greater, that is to say that it is more elongated from the center; but, on the other hand, the more the points are elongated from the center the more their chance of being achieved is weak. The probability for which an ellipse will be attained depends thus on two factors for which the one grows with the distance whereas the other diminishes. The maximum takes place for the more probable ellipse, the length of its rays is expressed by the number 0.85, in taking for unity the homologous rays of the probable ellipse.

The third ellipse, the probable ellipse is such that half of the projectiles fall inside of that ellipse and half outside.

Let us consider all the projectiles which fall on the vertical passing by the center of the target and above this center; if we take the mean of the distances of the points of impact of these projectiles to the center we obtain the mean écart in height. The corresponding point is the most elevated point of the mean ellipse. One would define likewise the mean horizontal écart and the mean écart following any direction. The rays of the mean ellipse are equal to those of the probable ellipse multiplied by 1.06.

There is one chance in four that the projectile falls outside of the ellipse of the last quartile. The rays of that ellipse are equal to those of the probable ellipse multiplied by 1.41.

It doesn't seem that one is able to give a simple demonstration that satisfies the law of écarts in shooting at a target.

The only good demonstration is that which rests on the hypothesis analogous to the hypothesis of the infinitesimal errors; it is complicated and painful.

Another demonstration, based on the extension of the postulate of Gauss, is simpler, but without great value. For these reasons that I developed previously, the hypothesis of Gauss cannot serve the unique principle of one theory.

If one assumes that the deviations of the shot in range and in direction are the same, the probable ellipse and all the analogous ellipses would reduce themselves to circles; the probability for which the projectile hit the target at a given point would only depend on the distance of the point to the center.

All would be simplified and the demonstration of the law of écarts would become very elementary. Unfortunately, such a supposition would be contrary to the reality and, from the rational point of view, it would much diminish the value of the theory.

What is important for us, is not to demonstrate the law, it is to thoroughly understand and to remember that the experiment always verifies it.

Does there exist a process permitting the equitable classification of the gunmen in a competition?

In order that the classification is made fairly, it is useful to make each competitor shoot a large number of projectiles.

One shot, in general, is characterized by five coefficients: two are related to systematic defects, three to fortuitous écarts. Let's suppose, in order to simplify, that there are not systematic defects and that the three characteristics of the fortuitous écarts are reduced to two (this amounts to admitting, as we have previously done, that the major axis of the ellipse of probability are vertical).

We can take for the characteristics of the probable écart in elevation and the probable écart in direction, or even, the mean écart in elevation and the mean écart in direction. These two écarts are proportional to the probable écarts.

If, for a gunman, the two écarts are smaller for one than the other, he must, equitably, be classed as first; but if one of the écarts is greater and the other smaller, there is doubt and the classification must result from a hypothesis on the relative importance of the écarts in elevation and in direction.

In other terms, if, for a gunman, the probable ellipse is inside of the probable ellipse of another gunman, the first is indisputably superior to the second; but if the ellipses intersect, there is doubt.

In practice, in order to simplify, one counts only the distance of the points of impact at the center of the target, one add these distances for each gunman and classifies the contestants according to the obtained sums.

Let us try to form a general idea and let's take a glance at that which is contained in this little book.

We have seen that the same law, that one calls the law of large numbers, presents itself in all important questions; it rules the écarts produced by the randomness in games, in the happening of events, in financial speculation, in the shooting at a target, in the measure of sizes and in many natural phenomenon. The knowledge of that law recognizes that, in our study, it is necessary to know the law and to remember it.

When at each proof, at each test, at each instant, randomness can produce some écarts which, with equal likelihood, are positive or negative, directed in one way or in the opposite way; in the long run, the resulting écarts follow the unique law, by one excessive simplicity, called *the law of large numbers.*

When, furthermore, the proofs, the tests, the successive instants are identified *a priori*, that is to say before the effect of randomness is differentiated, the resulting écarts are proportional to the square root of the number of tests or some tests or the square root of time.

That lone conclusion, which is very general and very elementary, suffices to show the usefulness of the calculation of probabilities; that calculation, which precedes the time of the philosophy and the science, which is at the same time very profound and very simple, which required much reflection and very few formulas should be studied by all the philosophers as by all the students, the one and the other would find there without doubt a very great interest and a very great charm; following the famous words of Laplace: "There is not a science more deserving of our meditations."

END

Credits

Cover image: Jean-Eugène Buland (1852–1926), *Le Tripot*. Oil on canvas, 63.5 × 109.22 cm. Original image in the public domain, edited by Kısmet Press, and republished under the same license. <https://commons.wikimedia.org/wiki/File:Jean-Eug%C3%A8ne_Buland_-_Le_Tripot_(1883).jpg>.

Curb brokers in Wall Street, New York City. 2 October 1920.Photoprint by Bain News Service. George Grantham Bain Collection (Library of Congress). Original image in the public domain, edited by Kısmet Press, and republished under the same license. <https://en.wikipedia.org/wiki/File:Curb_brokers_in_Wall_Street,_New_York_City,_1920.jpg> <https://www.loc.gov/pictures/item/92519195/>.

Broad Street and Curb Brokers, New York, 1898. Detroit Publishing Company postcards. The Miriam and Ira D. Wallach Division of Art, Prints and Photographs: Photography Collection. Original image in the public domain, edited by Kısmet Press, and republished under the same license. <https://digitalcollections.nypl.org/items/510d47da-370d-a3d9-e040-e00a18064a99> <https://commons.wikimedia.org/wiki/File:Broad_Street_and_Curb_Brokers,_New_York,_N._Y_(NYPL_b12647398-69764).tiff>.

Frontispiece: Edgar Degas, *At the Stock Exchange*, ca. 1879. Oil canvas, 100 cm × 82 cm. Musée d'Orsay, Paris. Original image in the public domain, edited by Kısmet Press, and republished under the same license. <https://commons.wikimedia.org/wiki/File:At_The_Stock_Exchange.jpg>. Details of painting can be found on the museum's website: <http://www.musee-orsay.fr/en/collections/works-in-focus/painting/commentaire_id/portraits-a-la-bourse-15775.html?tx_commentaire_pi1%5BpidLi%5D=509&tx_commentaire_pi1%5Bfrom%5D=841&cHash=0f98492659>.

Hungarian playing cards from 1930s. Credit to Takkk for photo. Original image licensed under a CC BY-SA 3.0 license. Edited by by Kısmet Press, and republished under the same license. Details of license here: <https://creativecommons.org/licenses/by-sa/3.0/deed.en>. <https://commons.wikimedia.org/wiki/File:The_old_cards_of_Hungary_-_1930's_year.jpg>.

Forgotten amidst the dusty archives of French technical literature lies a delightful little gem written by Louis Bachelier in 1914 entitled *Le Jeu, la chance et le hasard*, or *The Game, Luck and Randomness*. Popular in pre-World War I France, *Le Jeu...* offered the general reader of the time original insights into quantifiable patterns found in both casino games and financial markets; it was perhaps one of the first "how to get rich quick" books. But its popularity did not survive the early 20th century upheavals in Europe and it was never translated and published into English.

The book in hand, Sketches in Quantitative Finance, is the only known English translation of *Le Jeu, la chance et le hasard*. It attempts to be a technically precise, word for word, translation of Bachelier's work, preserving his original tone and his refreshingly readable (albeit quirky) style of writing. While this translation's entertainment value is enhanced by the frequent nuggets of knowledge of Bachelier's mathematical perspectives, perhaps the greater pleasure to the English reader will be to experience a freshly exposed artifact of scientific philosophy.

Edward Harding received degrees in geography, business, and information technology from Middlebury College, Dartmouth College and the University of Massachusetts respectively. He was a Professor of Finance at Plymouth State University in the US for over thirty years, teaching, writing and self-publishing several concise course textbooks in the field of quantitative finance. In this capacity he stumbled upon the work of Louis Bachelier and was captivated by the content: the math, the finance, the philosophy, and especially the slippery nature of randomness. He currently lives a quiet life with his wife in the woods of New Hampshire, USA.

www.ingramcontent.com/pod-product-compliance
Lightning Source LLC
Chambersburg PA
CBHW072255210326
41458CB00074B/1764
9780995671782